Quick Reference to Correctional Administration

Richard L. Phillips
and
John W. Roberts

AN ASPEN PUBLICATION®

Quick Reference to Correctional Administration

Richard L. Phillips
Criminal Justice Consultant
Geneva, Illinois

John W. Roberts
History Associates Incorporated
Rockville, Maryland

AN ASPEN PUBLICATION®
Aspen Publishers, Inc.
Gaithersburg, Maryland
2000

Library of Congress Cataloging-in-Publication Data

Phillips, Richard L.
Quick reference to correctional administration / Richard L. Phillips, John W. Roberts.
p. cm.
Includes bibliographical references (p.) and index.
ISBN 0-8342-1756-2 (alk. paper)
1. Prison administration—United States—Handbooks, manuals, etc.
2. Correctional institutions—United States—Administration—Handbooks, manuals, etc. I. Roberts, John W. (John Walter), 1954- II. Title.
HV9469.P49 1999
365'.068—dc21
99-045037

Copyright © 2000 by Aspen Publishers, Inc.
A Wolters Kluwer Company
www.aspenpublishers.com
All rights reserved.

Aspen Publishers, Inc., grants permission for photocopying for limited personal or internal use. This consent does not extend to other kinds of copying, such as copying for general distribution, for advertising or promotional purposes, for creating new collective works, or for resale. For information, address Aspen Publishers, Inc., Permissions Department, 200 Orchard Ridge Drive, Suite 200, Gaithersburg, Maryland 20878.

Orders: (800) 638-8437
Customer Service: (800) 234-1660

About Aspen Publishers • For more than 40 years, Aspen has been a leading professional publisher in a variety of disciplines. Aspen's vast information resources are available in both print and electronic formats. We are committed to providing the highest quality information available in the most appropriate format for our customers. Visit Aspen's Internet site for more information resources, directories, articles, and a searchable version of Aspen's full catalog, including the most recent publications: **www.aspenpublishers.com**
Aspen Publishers, Inc. • The hallmark of quality in publishing
Member of the worldwide Wolters Kluwer group.

Editorial Services: Lenda Hill
Library of Congress Catalog Card Number: 99-045037
ISBN: 0-8342-1756-2

Printed in the United States of America

1 2 3 4 5

Table of Contents

Preface v

Introduction vii

Chapter 1 — The Basics: Processing Inmates Into and Out of Prison...................... 1
    I. Receiving and Discharge Operations 1
    II. Admission and Orientation Program 17
    III. Pretrial Inmates 20
    IV. Classification 30
    V. Inmate Files............................. 37

Chapter 2 — Day-to-Day Supervision and Security Procedures 49
    I. Inmate Accountability..................... 49
    II. Searches................................ 53
    III. Posted Picture Files 58
    IV. Alcohol Testing 61
    V. Urine Surveillance 62
    VI. The Special Monitoring System 67
    VII. Special Housing 73
    VIII. Inmate Discipline 88
    IX. Use of Nonlethal Force and Application of Restraints 105
    X. Use of Firearms 115
    XI. Escorted Trips 118

XII. Voluntary Surrenders and Unescorted
     Transfers 123
XIII. Security Considerations Relating to
      Inmate Access to Computers 128

Chapter 3 — Inmate Entitlements **133**
I. Inmate Telephone Regulations 133
II. Inmate Correspondence 138
III. Visiting Regulations 149
IV. Inmate Grievance Procedures 156
V. Inmate Legal Activities 160
VI. News Media Contacts 166

Chapter 4 — Programs and Services **171**
I. Inmate Employment, Educational
   Programs, and Vocational Training 171
II. Recreational Activities 180
III. Religious Activities 185
IV. Inmate Organizations 189
V. Halfway House Placement 192
VI. Furloughs 198

Chapter 5 — Medical and Psychological Issues **205**
I. Infectious Disease Management 205
II. Suicide Prevention Program 219
III. Food from Outside Sources 226
IV. Inmate Hunger Strikes 227
V. Sexual Assault Prevention and
   Intervention 231

End Notes **239**

Index **247**

About the Authors **257**

PREFACE

Correctional management is an area of governmental operations that has received scant attention until the last several decades. Throughout the history of the United States, prisons have been out-of-sight, out-of-mind places—places most citizens had no interest in and no interest in knowing about. Other than through the occasional movie or media account of a prison escape or riot, most people in the United States knew very little about correctional issues.

That began to change in the 1970s, for a variety of reasons. Ordinary citizens began to hear about prisons in the context of events at places such as Attica and Santa Fe, where dramatic prison uprisings cost the lives of dozens of staff members and inmates. Other prominent media accounts told of the intervention of federal judges in entire prison systems, such as those in Texas and Alabama. By the mid-1990s, prisons were much more in the forefront, as the cost of a nationwide "war on drugs" became a reality. In prison systems across the country, construction and operating costs began to add up to budget-strangling levels. Suddenly, prisons became real to average, taxpaying Americans.

Today's prison administrators have a companion reality: much more is expected of them now than ever before. The public is aware of prisons. The public expects its prisons to be run effectively. The public expects its prisons to be run efficiently.

For years, the Federal Bureau of Prisons (BOP) and the prison systems of several states have assumed a leadership role in developing model correctional policies. But the "program statements" of the BOP are written in a very specific format, which conforms to federal publication standards that often makes the statements difficult to adapt for use by other agencies.

The authors of this reference, however, have taken key policies and procedures of the BOP and adapted them for general correctional use. These modifications make this book a highly valuable reference work for correctional agencies that need to update or upgrade their policy documents, or to expand a current body of policy into new areas. They also may be used to develop functional internal and external audit or program review formats for agencies that already have an in-house policy in place.

In addition, this reference could well be used in the academic world. So often, graduates of criminal justice and corrections programs enter the "real world" of corrections without a solid grounding in correctional practices. This volume provides the kind of practical, hands-on perspective on day-to-day prison life upon which most, if not all, new correctional workers could rely for a head start in a challenging and demanding career field.

Richard L. Phillips
John W. Roberts

INTRODUCTION

Corrections traditionally has been a profession that relied less on policy than on the instincts and common-sense experience of people who came up through the ranks to run prisons. And in the United States—because of its fragmented system of federal, state, and local corrections—management practices and structures have varied greatly. Some agencies developed detailed policies and procedures to guide staff, whereas others relied on more informal systems of management guidance.

As courts began to show an increased level of interest in corrections, beginning in the 1960s, the matter of correctional policy began to be ever more important to U.S. prison administrators. An agency without a policy to justify and guide a practice became an agency that could find itself in serious legal trouble. And as corrections rapidly expanded, beginning in the late 1980s, an influx of new staff and newly minted managers made it even more important that agencies have a solid policy base to direct the operations and actions of their staff.

The purpose of this volume is to provide readers—including staff of other public or private correctional agencies, scholars, students, and the general public—with a clearer view of correctional policy by presenting in a readable and easily accessible format the policies of the Federal Bureau of Prisons (BOP). The BOP has been one of the leaders in correctional professionalism and policy development, and many of its policies have been tested and validated in the courts. By removing references to federal statutes and regulations, and adopting a less formal style, the policies described in this book can be used as models for operations in nonfederal correctional agencies, as an informative course supplement for corrections classes, and as a valuable reference work for legal or other research.

No model correctional policy can be written in the abstract and then applied, unmodified. Each correctional system in the United States operates in a different statutory and historical context, a situation that certainly will influence the final shape of many prison policies. Likewise, collective bargaining agreements shape the form of actual prison operations in ways that may not be immediately evident to those outside the prison. And, as any correctional practitioner knows, the culture and customs of a given institution or correctional system are powerful and unique forces in the everyday operations of a prison.

For these and other reasons, no policy in this book can be applied intact and unmodified to an existing institution or even to a new one. But every one of these policies provides a legally sound and operationally practical way to approach an aspect of prison life. With careful review of the statutory, regulatory, contractual, and practical environment of a given agency or institution, these model policies can be a great aid to prison managers throughout the United States.

No correctional policy can stand alone. The best policy in the world—legally valid and beautifully written—is useless if it sits on the shelf in a binder. Staff have to be trained in it. Managers have to understand how it interacts with other aspects of the institution's operation. Supervisors have to be able to enforce it effectively and consistently. Line staff need to know how they are expected to act within policy. And inmates need to know key provisions of many important policies, as well. Some kind of policy compliance assessment system must be in place that includes a method for reviewing and revising policy, which can never be static in the face of both local conditions and the changing statutory and legal climate in which the agency operates. Without these ongoing activities and considerations, correctional policy is meaningless. But with those kinds of infrastructures in place, well-written correctional policy can be the backbone of an effective prison's operation.

The authors have benefited from the formidable assistance of Mr. Clair Cripe, retired general counsel of the BOP, who played a key role in reviewing the edited policies for legal validity and readability. His years of experience in the field of correctional legal work contributed greatly to the final product.

While these model policies were drawn from publicly available BOP program statements the content of this publication represents the views of the authors and not those of the BOP or the U.S. Department of Justice.

Richard L. Phillips
John W. Roberts

CHAPTER 1

The Basics: Processing Inmates Into and Out of Prison

I. RECEIVING AND DISCHARGE OPERATIONS

Overview

Controlling the movement of inmates and inmate property into and out of prison is a responsibility of paramount importance. Proper receiving and discharge operations must be in place to verify the identity of inmates coming or going, to ensure that inmates are being admitted or released appropriately and legally, to prevent the introduction of contraband, to maintain accountability over inmate property, and to maintain institution security.

The basic procedures to be followed when processing a new inmate into the facility should include: (1) verify the identification of the inmate; (2) verify commitment documentation; (3) conduct a pat search of the inmate; (4) separate inmate from property and funds; (5) notify other departments involved in the admissions process; (6) conduct a visual search of inmate (see Chapter 2, Part II); (7) take photographs and fingerprints; (8) inventory and/or issue property; and (9) ensure that social and medical screening are completed.

1. Processing Incoming Inmates

- The intake process begins when the staff is made aware of the pending arrival of an inmate. The staff determines which inmates are scheduled to enter or depart by consulting rosters and daily logs. A tracking system should be established to monitor volun-

tary surrender dates; if inmates do not arrive at the proper surrender date and time, the responsible referring agency should be contacted by telephone.
- The staff should prepare an admission folder for designated inmates, consisting of the designation packet and the receiving and discharge folder. An admission folder may be prepared for other types of commitments (such as pretrial, holdover, transfer, and special court commitment). The receiving and discharge folder for admission of designated inmates may include fingerprint cards, social and medical intake screening forms, an inmate account card, an admission and orientation handbook, and the acknowledgment forms, which should always include the inmate's name, institution or any register number, and the name of the institution.
- Officers of any agency who are committing prisoners are required to display official identification upon arrival. Credentials, badges, and other relevant paperwork should be closely inspected prior to allowing entry into the facility. The name of the officer delivering the inmate should appear on the commitment documentation and should coincide with the officer's personal identification. Whenever possible, the names of escorting officers should be obtained in advance of their arrival. Before officers are permitted into the facility, their ammunition and weapons must be secured by the receiving and discharge staff.
- The positive identification of each inmate is critical and must be carried out upon the inmate's arrival at the institution. It should be verified again when the inmate is removed from the holding area for intake processing. The identification process includes verbal questioning of the inmate as to name, date of birth, offense, sentencing court, and judge. Comparisons should be made against existing photographs, fingerprints, commitment papers, and physical description data.
- Commitment documentation must be verified carefully. A thorough review will ensure, for example, that inmates with short sentences and holdover inmates are not held past their release dates. Staff must review designation data for commitment dates, sentence length, and any special notations. The consolidated inmate file must be received with all transfers, and the completed accountability sheet must be attached. Commitment paperwork must be used to verify the delivery of inmates to the correct facility and to alert institution staff of significant facts regarding the inmate being committed. Exhibit 1–1 contains descriptions of some important commitment paperwork.

Exhibit 1-1 Examples of Commitment Paperwork

> *Judgment and Commitment Order.* This order must bear the signature of the judge and the seal of the court, and it should accompany the inmate on his or her initial arrival at the designated institution. The "return" section should be executed by the staff. If the certified copy has not been executed, staff at the designated institution must execute it upon the inmate's arrival. It is essential that the date of the original commitment to jail prior to sentencing is listed, as this will be used in determining the amount of credit to be allowed for time in jail or prior custody before sentencing. The name appearing on the Judgment and Commitment Order is the name to be used when admitting a sentenced inmate to the designated facility. It must appear on all paperwork and files exactly as it appears on the order; name changes shall only be accepted under a court order.
>
> *Parole Board Warrants.* The parole board issues violator warrants for the purpose of returning a parole violator to custody. The warrant contains the signature of a parole board member, the parole board's seal, and the date of issue. On the reverse is a space to execute the warrant. Following execution of the warrant, the original is returned to the parole board. Copies are retained by the institution in a judgment and commitment file as well as the inmate central file.
>
> *Transfer Orders.* This document establishes authority for intradepartmental transfers of inmates between facilities. It is usually signed by the warden of the transferring institution. It should be executed by receiving and discharge staff upon the inmate's arrival at the receiving institution. A transfer order is also used when inmates are transferred to halfway houses and outside medical facilities.
>
> *Writ Returns.* A copy of the original writ should accompany the inmate when he or she returns from a writ. It is to be executed by the returning agent to indicate that the writ has been satisfied. If the returning agent does not execute the writ, receiving and discharge staff should do so. The execution must include the time and date of the inmate's return, a statement that the writ was satisfied, and the signature of the staff member performing this function. If the writ is executed for the purpose of a prosecution, a Judgment and Commitment Order may also accompany the inmate. In the case of inmates returning from a release under the Interstate Agreement on Detainers, a copy of the Disposition of Charges may also accompany the inmate.
>
> *Court Orders.* Court orders for civil contempt, to direct confinement as a condition of parole or probation, to return to confinement from a court appearance or other temporary absence, or for pretrial services commitment may be used as commitment documentation. These orders must bear the signatures of the judge, the court seal, and the exact name of the inmate. They should be executed in the same manner as a Judgment and Commitment Order.
>
> *Remand of Prisoner Forms.* Remand of Prisoner Forms are used to accept custody of inmates from any agent or agency. This form can be used as commitment documentation on inmates being housed for safekeeping, overnight holdovers, or other similar situations.

Any inmate entering the institution from the community, from the court, or from a transfer where they may have come into contact with the public will be given a pat search prior to entering the institution. In addition, inmates will be visually searched (strip searched) and screened with a hand-held metal detector while in receiving and discharge. To initiate the search, the inmate should be placed in an area that cannot be viewed by other inmates or staff. The inmate is then instructed to disrobe, placing all clothing and personal property in a designated location. Normally, restraints will be removed before a visual search is conducted, and those restraints should be accounted for and removed immediately from the area. (For further directions concerning searches, consult Chapter 2, Part II).

- Staff must conduct a National Crime Information Center (NCIC) check on each inmate committed by a law enforcement officer before releasing the inmate from receiving and discharge to another area of the institution.

Photographs

- Identification photos must be taken of all incoming inmates. Photographs should depict inmates in full face, without glasses or head coverings; in profile; and wearing prescription glasses, if appropriate. An identification board with interchangeable letters and numbers, indicating the inmate's registration number and date of commitment, should be positioned immediately below the inmate's chin. New photographs should be taken whenever the inmate's physical appearance changes.

It is recommended that the Receiving and Discharge Department have a multiple-view camera, capable of taking four photos on one print. A digital photography system for photos may be used as an alternative. An additional camera should be available as a backup, should the regular camera become inoperable.

Photos should be attached to the receiving and discharge folder, crew kit cards, bed book cards, and other cards maintained at the facility. Extra photographs should be stored in the inmate's file for possible later use. At least one extra photo should be maintained in the inmate's judgment and commitment file for emergency use (such as escape fliers).

Fingerprints

- Three full sets of fingerprints should be taken on the initial commitment of designated inmates, including inmates being committed as release violators and escape returns. One full set of

fingerprints is placed in the judgment and commitment file, one set is forwarded to the state police, and one set is forwarded to the Federal Bureau of Investigation (FBI).
- Only one full set of fingerprints should be taken on the initial commitment of holdover or pretrial inmates. These should be stored with other appropriate file material.
- When an inmate is received as a transfer from another institution or is returned from a writ or a production under the Interstate Agreement on Detainers, and the FBI number is known, only the impression of the right thumb needs to be taken. This print is then compared with the card on file to verify identity; the new impression should replace previous fingerprint cards in the inmate's file (other than the initial full set of fingerprints, which should be retained).
- Inmate fingerprints should be taken on the official FBI Arrest Fingerprint Card (FD–249). The inmate should sign the card prior to printing, in order to avoid smearing. The inking surface should be elevated to a sufficient height to allow the inmate's forearm to assume a horizontal position when the fingers are being inked. A cleaning fluid and cloths or paper towels should be available to clean fingers before and after printing.
- Two types of impressions should be taken on the FBI fingerprint card. The upper ten prints should be taken individually, as rolled impressions—individually rolling each finger from side to side in order to obtain all available ridge details. The eight smaller impressions, located at the bottom of the card, are to be taken by simultaneously pressing down all the fingers. The thumbs are then printed alongside the other lower prints.
- The inmate should stand in front of and at a forearm's length from the inking plate. In taking rolled impressions, the side of the bulb of the finger is placed upon the inking plate and the finger is rolled. Each finger should be inked evenly from the tip to below the first joint. It is better to ink and print each finger separately, beginning with the right thumb, followed by the index, middle, ring, and little fingers. It is easier to print if the thumbs are rolled toward and the fingers rolled away from the center of the inmate's body. This process relieves strain and leaves the fingers relaxed upon the completion of rolling so that they may be lifted easily from the card without smudging. The officer taking prints should apply light pressure and maintain control of the inmate's hand during the process. The inmate should be cautioned to relax and refrain from trying to help by exerting pressure. It may be helpful

in getting inmates to relax their hands to tell them to look at some distant object and not at their hands.

It may be necessary to utilize special fingerprinting techniques if inmates have crippled fingers (bent, broken), deformities (webbed or extra fingers), missing fingers (due to birth defects or amputation), or problems associated with advanced age. In the case of inmates with crippled or deformed fingers, prints will be taken that are as good as possible. Only in those cases where the fingers are so badly bent or crippled that they are touching the palms and cannot be used is it permissible to refrain from attempting a print. In such cases, written notations should explain the absence of a print. The use of special inking devices similar to those used to fingerprint deceased inmates may be necessary to obtain clear, legible fingerprints from crippled fingers. Unusual circumstances should be indicated on notations entered on the fingerprint card, such as "finger missing at birth," "finger amputated," or "tip of finger amputated." Fingers with temporary disfigurements, such as wounds, blisters, and broken bones, should be printed as soon as possible after healing.

- Should one or more prints on a card be too indistinct to interpret accurately, the card will be rejected by the FBI and returned to the institution to be redone. The most common reasons for fingerprint card rejections by the FBI include: failure to roll finger fully from side to side, bulb of finger not completely inked, fingers allowed to slip, wrong type of ink used, failure to clean fingers or inking apparatus thoroughly, and excessive or insufficient amount of ink used.
- The following physical description and personal history data must be recorded on the FBI fingerprint card: sex, race, age, date of birth, place of birth, height, weight, scars and other marks, tattoos, eye color, and hair color.

Registration Numbers

- An inmate's registration number is assigned when the inmate first enters into the custody of the Department of Corrections; this number will be used throughout the inmate's entire period of incarceration anywhere within the prison system, and during any period of supervised release, for the current offense. Institutions housing pretrial inmates, safekeepers, or presentence admission cases are issued a block of registration numbers by headquarters;

these numbers are to be assigned to those inmates who are remanded without a previously assigned number. A name search should be performed before entering the inmate's name and number into the department of corrections' automated data system. Special care should be taken when searching for hyphenated names. If an inmate appears in the system under an old or incorrect registration number, staff must correct the problem immediately. An inmate should never be assigned a second registration number unless he or she is returning to prison under an entirely new conviction and sentence. All parole violators, escapees, study returnees after sentencing, and writ and furlough returnees should retain the registration number from their original commitment.

Other Issues

- New arrivals should have access to a shower prior to the issuance of intake clothing. The Receiving and Discharge Department should have an adequate supply of clothing and shoes for new commitments; these items should be available in various sizes and appropriate for the season and climate. Supervisory staff should be contacted for assistance and instruction if special arrangements must be made to obtain properly fitting clothes, if such are unavailable in the Receiving and Discharge Department.
- The staff member responsible for processing the new admissions should ensure that each inmate is cleared by the medical department and is provided with a social interview in accordance with policy on intake screening, prior to placement in general population. Therefore, appropriate intake staff must be notified as soon as new admissions are received in order to avoid delays in processing.
- All inmates should complete acknowledgment forms providing information on correspondence, authorization for disposition of funds, monitoring of telephone calls, visiting, notification in case of death or illness, and disposition of property in case of death or disability. A space is provided in each section of the form for the inmate's signature. Staff witnessing the inmate's completion of the form must also sign. If an inmate refuses to sign the form, the witnessing staff must so indicate in the space provided. Once completed, these acknowledgments will remain in effect throughout the inmate's confinement. The mailroom will be notified if the inmate elects not to have general mail opened and inspected.

The mailroom, cashier, and commissary supervisor will be notified if the inmate refuses to authorize the disposition of funds.
- An inmate commissary (or store) card will be prepared for all newly designated inmates. A label will be placed on the commissary card and signed by the inmate before the photograph is taken. All identifying data must be clearly reflected on the card and must not be altered in any way. The card is placed in the camera and a photo is taken of the inmate. A laminating machine should be used to seal the photo. The card will be given to the inmate after he or she signs a receipt for the card and agrees to pay for a replacement if it is lost or damaged. When an inmate is received in transfer with a commissary card that is still in good condition, that card should be returned to the inmate during the intake screening process.

2. Inmate Property

- Receiving and discharge staff are responsible for processing inmate property and ensuring that allowable items are given to the inmate and unauthorized items are mailed out of the institution. Sufficient storage space must be available in the Receiving and Discharge Department to secure inmate property.
- New admissions must be separated from their property, with the receiving and discharge staff completing an inmate personal property record form to inventory all property. All personal property must be thoroughly searched before being permitted into the secure area of the institution. All property received or purchased after initial commitment must also be documented on an inmate personal property record. Inmate personal property records should be retained in the receiving and discharge file, and a logbook—showing the inmate's name and registration number, the dates when property is received and issued, and the initials of the officer processing the property—should be maintained by the Receiving and Discharge Department to document all incoming personal property.
- In general, the only items of personal property that a new admission may bring into the institution are clothing (worn on person), a wedding ring, prescribed medicines and medical devices, currency or negotiable instruments, religious medals or medallions, a watch, prescription eyeglasses, and personal identification materials. In addition, an inmate's property may be accepted from

another institution, provided verification is made that the property has indeed been mailed from the facility in question. Packages received in the mail from other institutions must contain an inmate personal property record form, a copy of which should be forwarded to the receiving institution under separate cover for verification purposes. Selective Service cards, Social Security cards, driver's licenses, and other forms of identification must be taken from the inmate and stored in the inmate's central file.

Inmates may retain legal material, if it relates to ongoing litigation or if research material is unavailable at the institution. The General Counsel's Office for the Department of Corrections should be consulted to determine if specific legal materials are relevant to an inmate's case.

- During incarceration, inmates may receive packages from outside sources, with prior approval. An advance authorization form for receiving packages should be used to verify the contents of the package.
- Receiving and discharge staff should inventory and search all incoming inmate property. Inmates should be present, except in cases where the inmate is absent from the institution or where the inmate's presence would jeopardize safety and security. Inmates should sign a form indicating that the inventory information is accurate. Special procedures should be developed for handling the property of protected witnesses, to ensure confidentiality.

Although inmates should witness the inventory and search procedures, they should not have physical access to the property until it has been thoroughly searched and inventoried on the personal property record form. The inmate may not assist in the inventory or in the packing or unpacking of property.

Searched items should be kept separately from unsearched items and should be further separated according to disposition (those items to be given to the inmate, those to be stored, and those to be mailed out of the institution). Unauthorized items should be mailed home at government expense or, with the inmate's permission, donated or destroyed.

Special care should be given to detecting contraband concealed in inmate property. All clothing should be thoroughly inspected, giving special attention to pockets, seams, hat bills, hat bands, collars, waistbands, linings, cuffs, belts, and places where there is more than one layer of material. Unsealed commissary items, such as coffee, sugar, laundry soap, or other granular substances, must be poured into another container for inspection before being returned to the inmate. Shampoo, body oil, and similar items must be searched with a probe or pencil and scanned with a metal detector, if one is available. Special care should be used when inspecting any religious items, such as Bibles, religious headgear, medicine bags, etc., and the chaplain should be consulted to determine if articles are of religious significance. Shoes should be searched, with special attention given to detect hollowed out heels or cavities; the soles of shoes should be inspected, as they may conceal such flat items as hacksaw blades. Shoes should be scanned by a metal detector, if possible. Radios and books should be inspected. Staff should be alert when inspecting radios, as contraband tape players may be disguised to look like radios. Books must be opened, the pages searched, and bindings and covers (which are excellent hiding places for drugs and other contraband) carefully inspected. Pictures, picture frames, photographs, and photograph albums require special attention, as they are frequently used to conceal contraband.

Questionable items that cannot be thoroughly searched should be x-rayed, if possible. Items that cannot be thoroughly searched without being damaged or destroyed should not be permitted into the institution. Shipping containers and wrapping materials are to be treated as "hot trash" and not given to the inmate.

- There should be sufficient space in the receiving and discharge area to store property belonging to inmates on writ, property of incoming inmates, and property deemed discarded or abandoned. There must also be adequate space to store release clothing and clothing inmates need for writ or court appearances. Articles identified as "valuables" must be sealed in an envelope bearing the inmate's name and registration number and stored in a locked, fire-retardant vault, safe, or cabinet. Clothing should be stored in a secure room that is not accessible to inmates or unauthorized personnel. Copies of inmate personal property records documenting property in storage should be filed in a separate location.
- An inmate may be allowed to store property under the following circumstances: (a) the facility has locked deposit boxes available for the storage of valuables; (b) the inmate is admitted for short duration (such as a study or observation case), for civil contempt,

or for a short sentence; (c) the inmate is on pretrial status and is likely to have numerous court appearances; (d) the inmate is unable to provide a consignee or address to whom excess property can be mailed (in which case the property may be stored for up to 120 days—during which period receiving and discharge staff are responsible to determine at thirty-day intervals if a consignee or address has been obtained, and after which period the property will be identified as "abandoned" and disposed of in accordance with local procedures); or (e) the inmate is on holdover status.

- Upon release of an inmate to the community or to a halfway house, all property will be carried out of the institution by the inmate or mailed out by staff. Inmates must prove ownership of the property by producing an inmate personal property record, authorization form for receiving packages or property, commissary receipt, special purchase form, or other documentation. Staff should compare property against all property forms in the inmate's receiving and discharge file. The staff will search all property being mailed, prepare a personal property record for each box being mailed, and will pack and seal each box in the inmate's presence. Inmates should not assist in the packing process and must remain physically separated from the property. Each package will be logged in the outgoing property book, which is maintained in the receiving and discharge area. Matches, perishables, and currency or coins will not be mailed out. Unauthorized property may be confiscated. Once packaged, the property must be kept in a secure area until mailed, and mailing must occur within seventy-two hours of the inmate's departure.
- In addition, inmates may be authorized to mail property home during their confinement. The staff must document the mailing of outgoing personal mail, and inmates should incur the mailing costs. If property is returned as undeliverable, the staff will make at least two attempts to obtain a proper address. If the property is still undeliverable, the staff should return it to the inmate, unless it exceeds established limits on inmate property or is deemed contraband.
- All personal property belonging to an inmate being transferred to another institution will be mailed to the inmate's final destination by the originating institution. The staff will prepare mailing labels for this purpose. In the case of a holdover inmate being transported in a Department of Corrections vehicle, the personal property may be transported with the inmate.

All such property should be packed by staff as outlined above. The box should be labeled with the inmate's name, registration number, and final destination. An inventory will be recorded on an inmate personal property record form. The original and a copy of the inventory will be placed in the box for the receiving institution, a copy will be mailed to the receiving institution, a copy will be retained at the sending institution, and a copy will be given to the inmate. For groups of holdovers being transported with their property, an inmate property manifest will be prepared, listing the name, registration number, destination, and number of boxes for each holdover.

Medical centers should issue a list of approved property allowable for inmates temporarily transferred there for treatment. The staff will mail approved items to the medical center within seventy-two hours of the inmate's departure and place unauthorized items in the transferring institution's storage area until the inmate returns from the medical center.

- Property that is voluntarily abandoned by an inmate—that is, left behind by an inmate who has departed an institution, except through approved furlough—will be stored for thirty days after the inmate's departure or for thirty days after the close of any investigation. If the property is not claimed during that period, it may be utilized by the state or destroyed. Destruction of such property must be documented on an abandoned inmate property form.
- The personal property of a deceased inmate should be handled in accordance with the policy on escape and death notification. If the property remains unclaimed after two mailings, it should be considered abandoned and disposed of in accordance with local procedures.

3. Receiving and Discharge Facilities and Operations

The physical layout of the receiving and discharge area should include the following features:

- The area must be arranged in such a manner as to prevent searched and unsearched inmates from coming in contact with each other. New commitments and inmates being released must be kept separated at all times.

- There should be an adequate number of holding cells or other areas to provide for any necessary separation of inmates.
- A private area for conducting intake and medical screening must be available. There must also be a secure area for the storage of inmate personal property and court clothing. These areas must be inaccessible to other inmates and to unauthorized personnel.
- A sufficient number of lavatories and toilet facilities must be available to accommodate all inmates who may be processed in the area at any given time. Wash facilities must include an adequate number of sinks supplied with hot and cold water. Hand soap and towels or air blowers must be provided.

Holding cells must be searched before and after each occasion when inmates are placed in them. It is recommended that a logbook be maintained to record each cell search of this type. Windows in receiving and discharge holding cells should be screened, in order to deter the introduction of contraband. No areas holding inmates should contain false ceilings or furniture that is not firmly secured to the floor or wall. Holding cells should be situated so that staff may have visual contact of all cell occupants at all times. If the physical layout prevents such visual contact, alternatives—such as mirrors or cameras—should be employed.

Frequent security inspections of all receiving and discharge areas that are accessible to inmates must be conducted. Inspections should be carried out at varying times, in order to avoid setting definite patterns. Such inspections are designed to detect contraband, prevent escapes, maintain sanitation standards, and eliminate fire and safety hazards.

The Receiving and Discharge Department should conduct an "open house" at least twice a week to provide an opportunity for inmates to discuss receiving- and discharge-related issues with staff who perform those tasks. Receiving and discharge staff must also make regular visits to special housing units in order to answer questions. Questions submitted in writing from inmates should be answered in a timely and professional manner; a copy of each response should be placed in the inmate's receiving and discharge file.

It is strongly recommended that receiving and discharge officers wear a radio and a body alarm. It is also suggested that a set of handcuffs and a handcuff key be part of the regular issue. Keys for the receiving and discharge area should be "restricted" keys.

Ongoing training is essential to staff development, and it enhances the security and safety of staff and inmates. Supervisory personnel must schedule new employees for familiarization training in receiving and discharge procedures, and new employees should be assigned to work with experienced staff for a minimum of two weeks of on-the-job training. When new receiving and discharge procedures are published, supervisory staff must provide staff with training in those new procedures. Records office staff and receiving and discharge staff should be cross-trained in all functional areas of those departments, in order to provide coverage for staff vacancies and promote career development.

Receiving and discharge staff serve a critical role, as they are often the first point of staff contact for new inmates. Offenders are often committed while under the influence or withdrawing from the effects of drugs or alcohol. They may be psychologically unstable or angry and upset over their current circumstances. Therefore, it is imperative that staff be attentive and alert at all times. Receiving and discharge staff must exhibit a professional approach while performing their duties. It is extremely important for receiving and discharge staff to detect any unusual or volatile behavior on the part of incoming inmates and report this immediately to the appropriate institution staff.

Only staff should carry out the packaging and inventorying of inmate personal property, fingerprinting, filing forms, and taking photographs. Inmate workers must never be permitted to perform these functions. Inmate workers assigned to the Receiving and Discharge Department must not have any contact with inmates being processed in that area. Ordinarily, inmate workers in receiving and discharge should only be assigned to janitorial duties, although they may be assigned to tailoring duties if no other resources are available. Under no circumstances should inmate workers be assigned to clerical duties within the Receiving and Discharge Department.

4. Discharging (or Out-Processing) Inmates

- The staff must review all release paperwork to ensure that all required documentation is present. Any forms requiring the inmate's signature should be executed and the appropriate distribution made. Special monitoring clearance should always be obtained prior to releasing an inmate. When reviewing release paperwork, the staff should be particularly alert to the existence of detainers or warrants that are pending. Certain types of release will involve special types of documentation—such as a Furlough Application and Approval Form, the Escorted Trip Authorization

Form (in the case of local hospital day trips), a transfer order (in the case of an emergency overnight medical trip), bond releases, and others.
- Departments within the institution may request notification of an inmate's release. This notification ordinarily should be accomplished one working day prior to release. Inmates may bring their personal property to the Receiving and Discharge Department for pack-out at this time. Receiving and discharge staff must ensure that inmates listed on the notice of ensuing releases are properly processed. Coordination between the Receiving and Discharge Department and the records office is critical.
- On the day of release, the receiving and discharge staff will instruct the inmate to report to the Receiving and Discharge Department. Ample time should be allowed to complete the various release steps, including search; form completion; dress-out; receipt of medication, funds, and property; and final clearance. The staff must ensure that all processing is completed before the inmate's scheduled departure time.
- The staff must also ensure that all prisoners are properly identified prior to release. Identification is done by comparison of fingerprints and photographs, as well as verbal questioning of the inmate as to name, date of birth, offense, other commitment or personal information, and registration number. It is critical that, at the time of release, the inmate's thumbprint is taken and placed on the release authorization form. The thumbprint must be compared with the thumbprint in the judgment and commitment file to verify the identity of the inmate.
- A second identification verification should be made by a person designated by the warden. The person making the second identification must sign the release authorization.
- Receiving and discharge staff should conduct a visual search of the inmate. Clothing worn into receiving and discharge should be taken from the inmate, and the inmate should be dressed in clothing appropriate for the type of release. When escort and medical trips are processed through the Receiving and Discharge Department, special care should be taken to ensure that inmates are dressed in institution clothing and shoes. Inmates may have court or release clothing mailed to the institution. While awaiting departure, the inmate should be placed in a secure area, where he or she cannot have contact with unsearched inmates.
- Inmates being processed for temporary or permanent release may need to have access to prescription medicines. Medical personnel

will make any determinations regarding medication. Receiving and discharge staff must ensure that the individual being released has been provided with authorized medicines prior to departure.
- At the time of release, the inmate should be provided with a gratuity, as well as his or her personal funds. Funds generally are distributed by the cashier's office or through the control center.
- Inmates removed from the institution for court proceedings will be permitted to retain essential legal material, appropriate clothing for court purposes, personal hygiene items, prescription eyeglasses, dentures, prescribed medical devices, and prescribed medicines. Property removed from the facility with the inmate normally should fit into a 10" × 12" × 14" box. Other personal property should be stored at the institution. Inmates being released for court appearances may withdraw a reasonable amount from their commissary accounts. Staff may "close-out" the inmate's commissary account if it is known that the inmate will not return to the institution. Inmates not scheduled to return from court on the day of release must complete a disposition of mail form before leaving the institution.
- Inmates transferring to a local community medical facility ordinarily will be permitted to take prescription eyeglasses, dentures, prescribed medical devices, and prescribed medication. Other personal property and funds usually are not allowed.
- When an inmate is released to law enforcement agents, receiving and discharge staff should provide those agents with information on the inmate's criminal history, medical background, and record of behavior while incarcerated. It should be noted that law enforcement officials may arrive at the institution to pick up an inmate with other prisoners already in their custody for delivery to other destinations. If permitted under institution regulations, those officials may bring their prisoners into the institution to be secured while they are conducting their transactions. Staff should be cooperative and provide an area and supervision for those prisoners. Prisoners allowed into the facility temporarily must undergo the same shakedown as regular commitments.

Staff must exercise extreme caution when processing inmates for transfer to other law enforcement agencies. A thorough visual search must be conducted of the inmate and a hand-held metal detector should be run over the inmate's body just prior to departure. It is recommended that the receiving law enforcement officials conduct their own security search prior to accepting the inmate for departure.

Special procedures may be implemented for maximum-security inmates and other inmates having special security needs, including the rendering of assistance of medical staff to assist in the search of the inmate's person. Staff should escort the transporting officials and the departing inmates to the institution entrance to ensure that no contact is made with other inmates prior to departure.

Law enforcement officials taking custody of an inmate should be given appropriate documentation, including the In-Transit Information Form (completed by the unit staff), the inmate's medical record, and a release authorization. The Evidence of Agent's Authority to Act for Receiving State Form will be available to compare signatures of agents and to identify agents authorized to assume custody.

II. ADMISSION AND ORIENTATION PROGRAM

Overview

An inmate's initial impressions may be vital to his or her institutional adjustment, and attitudes formed early in an inmate's incarceration may influence his or her entire stay. Therefore, all newly committed inmates should participate in an admission and orientation program within the first few weeks of incarceration. An effective admission and orientation program provides the inmates with an overview of their rights and responsibilities, institution regulations, institution operations, program opportunities, and disciplinary practices. It also offers staff an opportunity to identify and assist inmates who may be experiencing unusual emotional stress or have other physical or mental health problems, hearing or learning disabilities, an inability to understand English, or other difficulties that could interfere with a successful adjustment to institutional life.

1. Eligible and Exempt Inmates

- All newly committed inmates and transfers from other prisons must participate in the admission and orientation program. Pretrial inmates and inmates in holdover status (en route to a different institution) need not participate in the program, although special procedures for orienting pretrial inmates are outlined in Part III.

2. Elements of the Program

- Lectures, group discussions, and visits to particular areas of the institution should include exposure to all programs relating to the specialized needs of the inmate, as well as exposure to various work assignments, education programs, recreation, health care, physical activities, and social interactions. The admission and orientation program should expose each inmate to as many work assignments as are consistent with the institution's security needs, and should be designed to determine the inmate's interests, aptitude, and experience in several areas of potential employment. When, for security or other significant reasons, it is not practicable for inmates to visit certain areas of the institution, audiovisual presentations may be substituted.

The schedule should ensure that inmates are involved in orientation programs during the day or early evening hours.

- Institution regulations, including discipline, visits, correspondence, and those policies governing other key aspects of prison life, should be a prominent feature of the admission and orientation program. Written materials should be prepared to supplement—but not replace—lectures and group discussions. The inmates should receive handouts to be kept for future reference.
- In addition to the institution's admission and orientation program, each unit manager is responsible for implementation of a unit inmate orientation program, appropriate to that unit's mission.

3. Location and Time Frames

- Each warden will establish procedures for the assignment of living quarters to newly committed inmates and will determine the appropriate location for the institution's admission and orientation program. The institution's schedule should be sufficiently flexible to accommodate the arrival of an inmate on any day of the week.
- The institution's admission and orientation program should be completed prior to the inmate's initial team meeting, which ordinarily would be held within four weeks of his or her arrival at the institution. Parole, mandatory release, and supervised release

violators are ordinarily classified within two weeks of arrival; admission and orientation programming for them should meet that time frame requirement.

4. Staff Responsibilities

- The admission and orientation coordinator should be an experienced program staff member, working under the direct supervision of a deputy warden or assistant warden, who will ensure that the goals and objectives of the admission and orientation program are being met.
- Each staff member assigned to present lectures as part of the admission and orientation program should develop an outline of the information they wish to include in their presentations and prepare a lesson plan for approval by the admission and orientation coordinator. The coordinator will maintain a file of those lesson plans. Staff will also develop written materials to supplement lectures and discussions.
- When a literacy problem prevents an inmate from understanding the admission and orientation information, it is the responsibility of staff members involved in the program to offer special assistance to that inmate. If an institution has a significant number of non–English-speaking inmates, admission and orientation information (including written materials) should be made available in a language understood by those inmates. During the admission and orientation process, staff must advise any inmate not fluent in English of the availability of translated documents.
- Any staff member involved in the admission and orientation program who believes that an inmate is experiencing significant emotional stress or other physical or mental health problems will notify the coordinator so that the inmate may be offered appropriate assistance.

5. Documentation

- To ensure that each newly admitted inmate completes the admission and orientation program, staff should develop an admission and orientation program checklist. As the inmate completes each phase of the program, staff should initial the checklist. When an inmate fails to complete a phase (which should occur only rarely),

the staff should initial the appropriate item on the checklist and indicate the reason why the phase was not completed. If possible, the inmate should be rescheduled as early as possible to complete the missed phase. The completed and signed checklist should be placed in the inmate's central file.
- In addition to an institutional checklist, the unit manager should be responsible for completing a checklist on the inmate's participation in the unit orientation program.

III. PRETRIAL INMATES

Overview

In addition to housing convicted inmates, correctional agencies typically are responsible for housing individuals who have been legally detained but for whom they have not received notifications of convictions. Commonly referred to as *pretrial inmates* or *pretrial detainees*, this class of inmates includes those who are awaiting trial, are in the process of being tried, or have been tried but are awaiting a verdict. It may also include those who have been convicted but have not yet been sentenced and those who have been committed for civil contempt of court, as deportable aliens, as material witnesses, or to undergo court-ordered mental evaluation or treatment.

Pretrial inmates are distinct from convicted inmates, who are individuals that a court has found guilty of an offense punishable by law. The term *convicted inmate*, however, refers *only* to an inmate's *current status*. An individual who has already completed service of any and all sentences previously imposed and is awaiting trial on a new offense should be considered a pretrial inmate rather than a convicted inmate. On the other hand, an inmate who is awaiting trial on one offense while still serving a sentence on a previous offense should be considered a convicted rather than a pretrial inmate.

Although pretrial inmates are subject to most of the same policies and procedures that are applicable to convicted inmates, there are certain special requirements pertaining to the care, custody, and control of pretrial inmates. Most notably, pretrial inmates should be housed separately from convicted inmates wherever it is practicable to do so. Also, pretrial inmates may not be compelled to perform work assignments other than housekeeping tasks in their own cells or community living areas.

1. Admission Procedures

Commitment papers on each incoming pretrial inmate must be verified. The staff should also obtain information from the prosecuting attorney or arresting officer concerning the inmate's behavior or offense severity, should question the committing law enforcement personnel about the inmate's security or medical precautions, and must enter this information on the Remand to Custody Form.

A search of the inmate's person should be conducted. The inmate should be photographed and fingerprinted. The inmate should be provided with clean clothing and personal hygiene items and should be given the opportunity for showering and hair care. The inmate's clothing and other personal possessions should be disposed of in accordance with regulations (see "Disposition of Inmate Property," below).

The pretrial inmate should be given copies of the institution's admission and orientation and an inmate rights and responsibilities forms, notified of institution guidelines concerning telephone calls, and asked to sign a Notice of Separation Form indicating that the inmate has been advised that he or she may or may not have contact with convicted inmates. If the inmate refuses to sign the Notice of Separation Form, then the staff should document this refusal on the form. Staff should also document on the orientation form that the inmate has received the appropriate pamphlets and instructions.

The pretrial inmate should be given the opportunity to waive the right not to work. The inmate may waive or rescind that waiver at any time.

2. Intake Screening and Assessment of Pretrial Inmates

Within forty-eight hours of their admission to the institution, pretrial inmates should undergo a thorough screening in order for staff to make an initial risk and needs assessment. This screening is essential to ensure pretrial inmates' safety and security and to minimize risk. It should also help staff identify seriously ill, violent, aggressive, escape-prone, or high-profile inmates who would require closer supervision or special housing. In addition, pretrial inmates should be screened again whenever they return to the institution following court appearances, because what happened in court may affect their status.

Intake screening should be carried out by unit managers, case managers, counselors, and physician assistants in most cases, although the warden may identify alternate staff, depending on institutional needs and capabilities. Unit staff conducting intake screening should seek information that may reflect on the inmate's behavior or offense severity, thereby helping to determine the inmate's security, medical, psychological, or other special needs. Unit staff should document their initial impressions and make a recommendation for housing to the unit manager.

An important way of obtaining information for the risk and needs assessment is for staff to interview the inmate directly. All interviews conducted as part of the intake screening process should be documented on an interview form developed for that purpose. Other information for this assessment may be obtained from the commitment documentation or from other law enforcement agencies. Staff must verify all information with the appropriate sources, and the information should be incorporated into the inmate's file for use during subsequent reviews. Sources for assessment information include the following:

- the Remand to Custody Form and accompanying booking information (both verbal and written)
- the medical and psychological screening and medical reports
- the pretrial services agencies (for bond information, prior record, and probation or parole report)
- the arresting agency (police department, sheriff's office, etc.)
- the prosecutor's office
- prior institution reports (reports on the inmate's conduct at state, federal, county, or local correctional facilities where he or she may have served sentences)
- the inmate interview
- the defense attorney
- the inmate's family

The following types of information are particularly important in determining the inmate's security requirements and health and psychological needs:

- separatee information (i.e., the inmate in question must be separated from other inmates)
- prior criminal history
- record of violence or threats against others

- escape or attempted escape from secure facilities, nonsecure facilities, or escort
- current offense or charge
- prior institutional adjustment
- age
- behavior and attitude during the intake screening
- special needs (psychological needs, medical needs, propensity toward suicide, speech problems, literacy problems)
- alcohol or substance abuse
- detainers or other pending charges
- bond information
- group affiliations (gangs, organized crime organizations, prior law enforcement connections; identifying marks such as tattoos)
- notoriety (high-profile cases in the news media)
- potential length of sentence, if the pretrial inmate should be convicted

A physician assistant or other medical staff designated by the warden should conduct an intake assessment to determine the inmate's medical and mental health status. Referral to the institution psychologist should be made as a matter of routine at this stage of the intake process, if a referral is requested by the inmate or if any staff member involved in the intake screening believes that a referral is appropriate. Procedures should be in place with appropriate law enforcement agencies and other agencies (including medical providers) to handle pretrial inmates who require medical care beyond that available at the institution.

Institution staff may commission the translation of documents, forms, and records into other languages when there are a significant number of inmates who speak or read only that language. In addition, important information contained in documents must be explained verbally to illiterate inmates.

3. Custody Level and Housing

- Ordinarily, pretrial inmates should be classified as "in" custody. Where circumstances warrant, however, staff may supervise a pretrial inmate according to procedures for other custody levels. Staff may consider a custody increase if verifiable information is available to justify such an increase. A memorandum approving this action, signed by the warden, must be maintained in the

inmate's file. A reduction in custody requires the approval at the headquarters or central office level by the assistant director or assistant commissioner in charge of operations. A written record, documenting the reasons for the reduction in custody level, should be maintained at headquarters and in the inmate's file.
- Housing assignments should be based largely upon information developed during the intake screening process, and decisions regarding housing assignments normally should by made by unit managers. Unless approved by the warden, in accordance with institution needs, the responsibility for making housing assignments should not be delegated to anyone below the level of unit manager.
- As a general rule, pretrial inmates should be housed separately from convicted inmates (provided that the physical layout of the institution permits this). Pretrial inmates who present risks to the security and orderly operations of an institution should be housed in a unit where appropriate security is provided. Where practicable, separation from convicted inmates should still be maintained. At subsequent pretrial inmate reviews, however, the staff should consider whether placement in less secure housing is appropriate.
- Deportable aliens who remain in custody immediately after completing service of a sentence for a felony conviction and are awaiting deportation do not need to be housed in a pretrial unit separate from convicted inmates. Program involvement and work assignments (other than housekeeping tasks) remain voluntary.
- The status of pretrial inmates must be reviewed whenever they return from a court appearance. Inmates in pretrial status who have pleaded guilty or been found guilty in a court of law immediately cease to be regarded as pretrial and no longer need to be separated from convicted inmates. They should be placed in "holdover" status, pending sentencing and initial designation.

4. Status Reviews

- Unit staff appointed by the warden must conduct regular reviews of the status of each pretrial inmate. A status review team should be composed of at least two unit staff members, ordinarily selected from among the unit manager, case manager, and correctional counselor. The unit manager is responsible for scheduling pretrial

The Basics: Processing Inmates Into and Out of Prison 25

inmate status reviews and for determining whether the inmate has been found guilty by the court since the last review.
- Each pretrial inmate should be scheduled for an initial status review by the unit team within twenty-one calendar days of his or her arrival at the institution, and subsequent reviews should be conducted at least every ninety days thereafter. The initial and subsequent reviews should assess all factors relating to the inmate's detention, including the practicability of separation from convicted inmates.
- The inmate should be notified at least forty-eight hours prior to a scheduled review and is expected to attend reviews. If the inmate refuses to appear at a status review, then staff should document the inmate's refusal in the record of the meeting and should also note the reason for the refusal (if known). All pretrial inmate status reviews must be documented in the inmate's file.

Any pretrial population may include high-security, high-profile inmates who could pose significant threats to themselves, other inmates, staff, or the outside community. The need to identify and monitor those cases regularly is of paramount importance. Unit staff should be aware of each of these inmates' court appearances and be ready to facilitate their transfer to another institution upon sentencing and designation.

Proper status reviews require complete information. To ensure that information about a pretrial inmate is as complete as possible, the institution should coordinate with the prosecutor's office, the office of the clerk of court, the probation or parole agency, the state police, and other criminal justice agencies. Contact between the institution and the prosecutor's office and local law enforcement agencies is particularly important to ensure timely notification when an inmate has been found guilty and may therefore be removed from pretrial status. All contacts of this nature with other agencies should be documented in the inmate's file.

5. Disposition of Inmate Property

- A pretrial inmate may retain personal property as authorized for convicted inmates housed in administrative segregation units. Institutional procedures for the handling of pretrial inmate property should be consistent with any instructions from the court.

Institution staff will be responsible for proper storage and accountability of a pretrial inmate's property. Inmate property that is of potential use to law enforcement agencies may be released to them with a court order, authorization of the prosecutor's office, or consent of the inmate.
- The institution may store the pretrial inmate's unauthorized property until the individual is released, transferred to another detention facility, or sentenced and designated to another facility. Procedures for the storage of unauthorized personal property should include provision for secure storage of valuables and other small items; the use of property inventory forms, to be signed by the detainee; limitations on access to property storage areas; and provisions for property to be turned over to third parties as designated by the pretrial inmate or returned to the pretrial inmate upon release. Unauthorized property that cannot be stored at the institution should be mailed, at government expense, to an address furnished by the inmate.
- A pretrial inmate may be permitted to obtain clothing for court appearances. When the institution provides the pretrial inmate with court clothing, the court should be contacted to provide information on the least expensive or elaborate clothing requirements deemed appropriate for courtroom appearances.
- Pretrial inmates may deposit funds in institution commissary accounts and should be subject to general regulations governing inmate access to currency, inmate commissary accounts, and the deposit and release of funds. Staff must advise pretrial inmates of those policies. The institution should establish procedures governing the return of funds and personal property to pretrial inmates released at a time outside normal business hours.

6. Discipline

Each pretrial inmate is subject to the correctional agency's policy on inmate discipline. The institution should advise the court of repeated or serious disruptive behavior by a pretrial inmate. The institution should also advise the prosecutor's office, the inmate's attorney of record, and the probation or parole officer assigned to prepare the presentence investigation report whenever a pretrial inmate violates any provision of the policy on inmate discipline.

7. Inmate Access to Legal Resources

- Pretrial inmates should have the opportunity to meet with their attorneys of record on a 7–day-a-week basis. Pretrial inmate attorney visits may be conducted at times other than regular visiting hours, with the approval of the warden. The warden should establish hours for attorney visiting and communicate them to the local legal community.
- Pretrial inmates should be permitted to telephone their attorneys of record as often as institution resources permit.
- Pretrial inmates must have access to legal materials housed at the institution. A basic law library should be established for each pretrial housing unit if inmates in that unit are not permitted to go to the institution's main law library. Procedures for obtaining legal materials not contained in a basic law library, but contained in the institution's main law library, should be established.

8. Medical and Mental Health

- The institution should provide pretrial inmates with the same level of basic medical care, dental care, and mental health care that it provides to convicted inmates.
- Staff should advise the court of any medication given to a pretrial inmate that may alter that inmate's courtroom behavior. The court should be notified of any psychotropic drugs administered to a pretrial inmate or an inmate undergoing court-ordered mental health evaluation. Except in event of an emergency requiring the immediate administration of such medication, the court should be notified before the inmate receives the medication. The prosecutor's office and the inmate's attorney of record should also be notified if the inmate is given medications that may alter his or her courtroom behavior. Judgment of whether a particular medication may alter an inmate's courtroom behavior must be made by a qualified health care professional, and the notification responsibility may not be delegated below the department head level.
- If an inmate should become seriously ill or should die, staff must notify the court, the prosecutor's office, the inmate's attorney of record, and the inmate's designated family member or next-of-kin.

9. Counseling

- When consistent with institution security and good order, pretrial inmates may be allowed the opportunity to participate in counseling activities, such as drug treatment, with convicted inmates. Pretrial inmates who are not permitted to receive counseling services with convicted inmates should be granted access to other counseling services.

10. Release of Information

Requests for information about particular pretrial inmates may be received from members of the news media or others. Public information officers at the institution should only indicate whether a person is or is not confined at the institution. Requests for any additional information about the pretrial inmate should be referred to the prosecutor's office.

11. Program Activities

- Unless a pretrial inmate signs a waiver of his or her right not to work, the warden cannot require that inmate to perform any job assignment other than housekeeping tasks in his or her own cell or community living area.
- Pretrial inmates are not eligible for furloughs and, except by court order, may not participate in any programs in the outside community. If an emergency arises where it might be deemed appropriate for an inmate to leave the institution, staff should facilitate contact between the inmate and his or her attorney of record, who in turn may seek a court order for an escorted trip or release from custody. Staff should also notify the prosecutor's office and the appropriate law enforcement agencies of any such emergency and document these notifications in the inmate's file. The institution should establish procedures with the appropriate law enforcement agencies to handle cases where escorts of pretrial inmates are necessary.
- Pretrial inmates may be allowed to participate in religious programs with convicted inmates, but only if doing so is consistent with institution security and good order. Pretrial inmates who are not able to participate in religious programs with convicted in-

mates must have access to other religious programs. Pretrial inmates should be allowed to possess one Bible or other religious books and may be visited by religious advisors from the outside community.
- A pretrial inmate may request permission to marry, in accordance with the correctional agency's policy for convicted inmates. Staff should advise the court and the prosecutor's office of the marriage request, and, in the case of aliens, should also advise the U.S. Immigration and Naturalization Service. Staff contacts with these agencies should be documented in the inmate's file. Ordinarily, a marriage should be allowed only if it has been approved by the court and the prosecutor's office.
- Pretrial inmates may have access to the institution's educational program, if it is consistent with institution security and good order. In addition, they may participate in correspondence courses and self-study courses. The staff may arrange for pretrial inmates to receive educational assistance through contract personnel or volunteers from the outside community. Pretrial inmates should be permitted access to the institution's inmate library and may check out books on the same basis as convicted inmates.
- When consistent with institution security and good order, pretrial inmates may be allowed the opportunity to participate with convicted inmates in some recreational activities. Ordinarily, pretrial inmates may participate in recreation together. Except in instances where compelling and documented safety or security reasons make it inappropriate to do so, pretrial inmates should be allowed one hour of outside recreation daily (weather permitting) and two hours of indoor recreation daily. Pretrial inmates housed in administrative segregation or disciplinary detention, however, may participate in recreation only as permitted under the correctional agency's policy on inmate discipline.

12. Visiting Privileges

- Pretrial inmates may receive visits in accordance with the correctional agency's policies and local institutional guidelines on visiting privileges. At a minimum, pretrial inmates should be approved for visits from immediate family members. These persons would include an inmate's mother, father, stepparents, foster parents, siblings, spouse, grandparents, grandchildren, children, and stepchildren.

- If a pretrial inmate does not have a spouse but has been cohabiting with an individual, and there is evidence that the relationship was similar to a spousal relationship, that individual should be approved for visiting. The preexisting relationship must be documented (for example, through names on a lease, birth certificates of children, or a common address on a driver's license) and subject to the normal screening procedures prior to approval.

IV. CLASSIFICATION

Overview

Classification contributes to institutional safety and efficiency by seeking to ensure that the appropriate security and program needs of inmates are identified. Through the classification process, inmates are assigned to housing commensurate with their security needs, program recommendations may be made, and institutional adjustment can be monitored and evaluated. To accomplish these ends, each newly committed inmate should be assigned to a classification team upon entering the institution to which he or she has been designated for service of sentence, should be classified within four weeks of their admission, and should be subject to subsequent program reviews at regular intervals.

1. Classification Team

- Each department within the institution should have the opportunity to participate in the classification process. At institutions that operate under unit management, each classification team should include the unit manager, a case manager, and a counselor. At institutions that do not operate under unit management, each team should include a case manager, a counselor, and at least one other staff member. An education advisor and a psychology services representative should also be members of the classification team.
- The education advisor serves as the unit classification team's consultant and expert in all educational, recreational, and vocational training matters. The education advisor may be assigned to one or more unit teams and consequently may be unable to attend initial classification or program review meetings. While it

is preferred that the education advisor attend such meetings, attendance is not mandatory and may be waived with the warden's approval. In lieu of attendance at any initial classification meeting or subsequent program review meeting, the education advisor should furnish the unit classification team with a written report outlining the inmate's progress or needs in the areas of education, recreation, and vocational training matters.
- The unit psychologist is not required to serve as a regular member of the unit classification team. However, unit psychologists are responsible for providing the unit classification team with written psychological reports on inmates scheduled for initial classifications or program reviews. Moreover, the unit psychologist should attend meetings on cases presenting special mental health challenges or concerns.
- The medical department is responsible for ensuring that all medical duty status assignments are accurate and that current medical information is forwarded to the unit classification team at least one week prior to a classification or program review meeting. Counselors or other unit staff should contact the hospital staff as necessary to ensure that inmate information system assignments are current and correct.
- During the initial classification, the case manager must review the inmate's central file thoroughly to determine if the inmate has any physical or mental disabilities. If any such disabilities are known or suspected, the case manager should notify the hospital administrator or chief of mental health services, by completing an inmate disability report. The case manager should also review the inmate's disability status or needs at scheduled program reviews and make referrals as necessary.
- On an individual basis, each member of the classification team should interview the newly arrived inmate within five working days of the inmate's assignment to that team. At these interviews, all unit classification team members should familiarize themselves with the inmate and with any areas of concern identified during the intake screening and orientation process. The unit manager is responsible for establishing procedures to ensure that these interviews take place within the five-day time period. When classification occurs in conjunction with the unit admission and orientation program, the prompt interview of inmates by unit team members identified on the unit admission and orientation checklist can facilitate compliance with the five-day interview requirement.

2. Initial Classification and Subsequent Program Reviews

- There are two types of unit classification team meetings: initial classification and program reviews. Initial classification occurs within four weeks of an inmate's arrival at the designated facility, following sentencing and commitment. All subsequent reviews of an inmate's status or progress are considered program reviews. When an inmate is redesignated, a program review must be held within four weeks of the inmate's arrival at the new facility.
- Unit managers monitor the scheduling of unit classification meetings. Therefore, unit managers are responsible for ensuring that case managers meet initial classification time requirements.
- An inmate's return to custody from the community requires a punctual reassessment of the inmate's program and security needs, especially in cases where allegations of new criminal conduct or technical violations of supervision may have occurred. Thus, an inmate who returns to an institution as a parole violator, mandatory release violator, or supervised release violator should be scheduled for an initial classification within two weeks of arrival. All subsequent unit classification team meetings for inmates returning to custody under these circumstances are considered program reviews.
- Former study cases (court-ordered forensic or competency studies) should not be scheduled for initial classification until the inmate's central file is available.
- Staff should conduct a program review for each inmate at least once every 180 days. When an inmate is within twelve months of his or her projected release date, a program review should be conducted at least once every 90 days. Program reviews may be scheduled more frequently than every 90 or 180 days at the discretion of the unit classification team.
- In some cases, a program review date may occur while an inmate is away from the institution or in the custody of other law enforcement agencies. Under such circumstances, the unit classification team should hold a program review meeting within two weeks of the inmate's return. Otherwise, the inmate's next program review will occur as scheduled.
- It is possible that initial classifications or program reviews may be delayed for inmates who are being held in special housing units at the time their unit classification team meetings are scheduled to take place. In those cases, the meetings should be held as soon as

practicable—ordinarily no later than two weeks after a regularly scheduled meeting was supposed to occur.
- Unit classification team members should review all relevant information before the initial classification or program review meeting. The various team members have differing responsibilities, as discussed below.

The *unit manager* serves as chair of the classification team, monitors the scheduling of an inmate's appearance at classification or program review meetings, and ensures appropriate documentation exists when an inmate declines to appear. Following team meetings, the unit manager must ascertain that all inmate information system assignments are current and accurate; that all inmate forms are correct, complete, and filed in the appropriate section of the inmate central file; and that all court orders are reviewed to ensure that any judicial recommendations requiring unit classification team action have been addressed. The unit manager is also responsible for obtaining the inmate's commissary record in advance of all program reviews; the commissary record will show the inmate's account balance and purchasing activity since the prior meeting. This can be used to help review and evaluate the inmate's financial responsibility program status.

The *case manager* schedules the inmate's appearance before the unit classification team. During the meeting, the case manager should provide an oral summary of the inmate's current offense, prior record, social situation, and security/custody classification. The case manager also provides a synopsis, as appropriate, of inmate special monitoring concerns, financial responsibility program obligations, victim/witness issues, and special programming considerations (drug abuse program, mental health counseling, etc.).

The *education advisor* provides the unit classification team with consultation and expertise in all matters relating to education, release preparation, recreation, and vocational training matters. Education recommendations or requirements should include target dates for program completion and standards of measurement that may be used to assess the inmate's participation in courses or programs.

The *correctional counselor* should present an oral summary of the inmate's work performance, general adjustment, living quarters sanitation rating, and other information relevant to the inmate's confinement. The counselor should obtain this information from unit officers, work detail supervisors, and other staff who have contact with the inmate. When the education advisor is unable to attend a unit classification team meeting, the counselor should present an oral summary of the inmate's education test results, recommended educational program needs, and progress toward completion of education and other applicable release readiness programs.

The *unit psychologist* should submit a psychological report on the inmate upon the request of any unit staff member or if he or she believes that a separate report on the inmate's mental health status is warranted.

The staff must notify an inmate that he or she has been scheduled to appear before the unit classification team, either for an initial classification or for a program review, at lease forty-eight hours in advance. The forty-eight-hour notification requirement may be waived by the inmate, in writing.

- The inmate is expected to attend the initial classification meeting. If the inmate refuses to appear at this meeting, the staff should document the inmate's refusal in the official record of the meeting and, if known, the reasons for the refusal.
- An inmate may elect not to attend subsequent program reviews, but should indicate this intent in writing at least twenty-four hours before the scheduled team meeting. If an inmate does not attend a program review meeting and fails to provide a signed statement to that effect, then the staff should document the inmate's refusal to appear in the program review report and, if known, the reasons for the refusal. In such cases, a copy of the program review report should be forwarded to the inmate. The inmate is responsible for becoming aware of, and will be held accountable for, the unit classification team's actions.
- The staff should complete a program review report at the inmate's initial classification. This report should include information on the inmate's needs, as identified by the team classification committee, and should outline a correctional program designed to meet those needs. The program review report should be signed by the unit manager and the inmate. A copy must be placed in the inmate's central file and a copy must be provided to the inmate.

In the program review report, the correctional program should be stated in measurable terms, establishing time limits, performance levels, and specific, expected program accomplishments. The correctional program should include a work assignment for each sentenced inmate who is physically and mentally able. Except for the work assignment, the inmate's participation in the prescribed program is voluntary (unless required by policies of the correctional agency, by court order, or by statute).

At subsequent program reviews, the staff should document the inmate's progress in meeting goals laid out in the correctional programs. Changes in the inmate's correctional program that are approved as part of program reviews should be set forth in the program review report in the same manner, specifying time limits, performance levels, and program accomplishments.

- When providing the inmate with a copy of the program review report, the staff should obtain the inmate's signature verifying receipt of the report. If the inmate refuses to sign for a copy of the report, staff witnessing the refusal should place a signed statement to this effect in the report and in the inmate's central file.
- A staff summary, prepared in memorandum form and signed by the case manager and the unit manager, is required for inmates for whom no presentence investigation report is available, for inmates who are serving a period of study and observation, and for inmates who are foreign nationals and have applied for transfer to their native country under provisions of a treaty transfer program. In such cases, the staff summary should be completed within five working days of the initial classification or before the completion of the study and observation case. It would include information on the inmate's current offense and prior record, status of pending charges, level of education, marital history, substance abuse history, physical health status and history, mental health status, and community resources potentially available to the inmate. A copy of the staff summary should be provided to the inmate upon the inmate's request.
- The unit classification team should conduct custody classification reviews in conjunction with program reviews when possible. A team docket should be prepared for each meeting of the unit classification team (whether for initial classifications or program reviews), listing the name and registration number of each inmate scheduled for appearance and the date and time of the meeting. These dockets should be posted conspicuously in the unit at lease forty-eight hours prior to the meeting. Following the team meeting, team actions should be documented on the docket sheet and signed by all staff present at the meeting. Docket sheets should be retained by the unit manager for at least one year after the meeting occurs.

3. Unscheduled Reviews

Unscheduled program reviews may take place in advance of regularly scheduled program reviews upon request of the inmate or a member of the unit classification team, with the concurrence of the chair of the unit classification team (usually the unit manager).

4. Appeals Procedure

Inmates may follow regular inmate grievance procedures to appeal any decision made at the initial classification or at a program review.

5. Study and Observation Cases

Inmates committed to an institution for purposes of study and observation are exempt from classification and program review procedures, apart from the requirement that the unit classification team prepare a staff summary (as noted in "Initial Classification and Subsequent Program Reviews," above).

6. Detainers

The existence of a detainer ordinarily should not affect the inmate's program. An exception may occur where the program is contingent upon a specific issue (such as custody) that is affected by the detainer.

7. Staff Training

The case management coordinator should incorporate the policy on classification and program reviews into regular case manager and unit secretary training sessions. Unit managers should ensure that each member of the unit classification team is familiar with all requirements of the classification and program review policy. The case management coordinator should attend meetings of each unit classification at least once every six months, in order to monitor case management consistency.

V. INMATE FILES

Overview

Maintaining complete, accurate, and well-organized inmate files contributes to effective inmate management, ensures better administrative efficiency, and provides important legal protections. Without a properly maintained file on each inmate, it would be virtually impossible to develop, implement, or monitor an inmate's custody status or correctional program or to resolve questions regarding disciplinary measures taken against an inmate, release dates, or even the institution's authority to incarcerate a particular inmate.

Inmate files typically are broken down into an inmate central file, a privacy file, a parole board file, and a sentence and commitment file. The nature and contents of each are outlined below.

In most jurisdictions, inmate files maintained by a publicly operated (as opposed to a privatized) prison or jail are subject to laws and regulations affecting government records. These laws and regulations vary from jurisdiction to jurisdiction, and always should be consulted before destroying records, retiring them to an archival facility, or opening any portion of them to public scrutiny. At a minimum, these laws and regulations should provide guidance on developing a legally defensible "records schedule" for inmate records, which would set forth the number of years an institution should hold inmate files on site before destroying them or retiring them to an archival facility. Not only is this necessary to ensure that the institution or the correctional agency is in compliance with statutes for handling government records, but it is also a useful managerial technique for ensuring that files may be retrieved in an efficient fashion and that older files are not kept on site indefinitely, thereby creating storage problems.

1. Assignment of Responsibilities for Handling Inmate Records

- Responsibilities for creating and maintaining inmate files are shared by the unit and the records office. It is the responsibility of the unit to create the central file, the privacy folder, and any necessary files for the parole board; to monitor the filing of documents within the inmate central file; and to retain custody of

the file as long as the inmate is assigned to that unit. The unit is also responsible for ensuring that the inmate file is properly secured. The records office, meanwhile, is responsible for secure filing of the commitment papers and other court-related and sentencing documents. The records office also must coordinate the movement of central files and central file material between units. This office also handles the transfer of files to other institutions or offices within the correctional agency and is in charge of retiring and retrieving inactive files.
- In addition, the correctional agency's headquarters or central office may maintain a duplicate central file on each inmate within the system and may provide guidance and supervision to records offices at individual institutions.
- Procedures and responsibilities for creating and handling files related to special witnesses, witness protection cases, or other special monitoring cases may differ from procedures for standard inmate files, due to the exceptional sensitivity of those cases. Specific instructions for file keeping may be contained in agency policies and regulations for the special monitoring system or special witness units.

2. Creation of the Inmate Central File

- Unit secretaries should review daily logs for inmate assignments to their units. For each new admission being processed into a unit at the designated institution, the unit secretary should create an inmate central file and privacy folder as part of the admission process.

New central files ordinarily do not need to be set up for inmates who are not new admissions to the correctional agency, but were previously classified and are being transferred to a unit from elsewhere in the system, because their existing files would be transferred to the units to which they have been newly assigned. Similarly, new central files ordinarily do not have to be set up for former inmates who have been recommitted under the same sentence, because their inactive files should be reactivated.

- Each inmate central file should be stored in an appropriate folder, along with pressure-sensitive labels. Tattered or torn file folders should be replaced.

3. Components of the Central File

- The inmate central file may be broken down into as many as six sections, thereby permitting the organization of filed material by topic. These sections are described in Exhibit 1–2. Inmate grievance forms *should not be maintained in the central file*; rather, they should be kept in a separate inmate grievance file.

4. Privacy Folder

A privacy folder should be attached to the top of Section 5 of the inmate central file. Documents that, by law, are exempt from disclosure to the inmate should be placed in the privacy folder. Such documents would include special monitoring forms; correspondence or any other documentation regarding special monitoring issues; victim/witness notification requests from courts, victims, or witnesses; initial victim/witness notification letters and all other notification documents; nondisclosable presentence information and court materials; reports on study and observation cases; and reports on confidential investigations.

5. Parole Board Files

- To facilitate the sharing of information between the correctional agency and the parole board, the unit secretary should create a parole board file at the time of admission for each commitment who is eligible for parole and who eventually will be released to the supervision of the parole board. In creating and maintaining parole board files, institutions should observe parole board rules and regulations. Each parole board file is held at the institution where the inmate is confined. It should be shipped by certified mail to the parole board office prior to parole hearings and then returned to the institution following the hearing. Upon the inmate's release from confinement, the parole board file should be sent to the supervising parole board office.
- Parole board files should contain the following documentation: (a) the sentence computation record; (b) FBI or police fingerprint records; (c) a copy of Judgment and Commitment Order; (d) the presentence report, if disclosable; (e) the report of prosecutor's office; (f) classification material, including copy of sentence data

Exhibit 1–2 The Sections of the Central File

> *Section 1: Sentence Data/Detainers/Warrants.* Contains: (a) special monitoring system tracking card; (b) sentence data summary, with mug shot attached; (c) inmate information system sentence computation record; (d) copies of Judgment and Commitment Order and papers; (e) good-time computation forms; (f) detainer correspondence; (g) Interstate Agreement on Detainer Forms; (h) FBI fingerprint record and corresponding state or local police records; and (i) publicly disclosable correspondence from the sentencing judge, prosecutor, or representatives of other criminal justice agencies.
>
> *Section 2: Classification and Parole Materials.* Contains: (a) work assignment record; (b) in-transit data form (most recent only); (c) copy of transfer orders; (d) copy of redesignation approval or Inmate Information System clearance data (no separatees listed); (e) requests for redesignation; (f) current/exception custody classification form; (g) all parole board forms (in chronological order), including parole board appeals, notices of parole board actions, waivers (of notice, representation, or disclosure), notices of hearing, parole applications and waivers, background statements of inmates, and attorney/witness election forms; (h) all related correspondence with the parole board, including parole violation warrant applications and probation or parole packets; (i) progress reports and halfway house terminal reports; (j) classification material, including program review forms and staff summaries; (k) clemency report; (l) request for presentence investigation report; (m) institutional designation form; (n) presentence investigation report (only if it is considered disclosable) and probation or parole violation report; and (o) correspondence relating to the presentence report (disclosable information only).
>
> *Section 3: Mail, Visits, Property, etc.* Contains: (a) approved visiting lists and supporting materials (excluding NCIC or police record checks from disclosable portion of file); (b) telephone information, including request for telephone privilege form; (c) inmate personal property records, including property confiscation forms; (d) commissary issue card form; (e) inmate package records; (f) reports on inmate injuries; (g) receipt of safety regulations; (h) receipt of rights and responsibilities documentation; (i) institution AIDS training record; (j) institution admission and orientation program materials; (k) unit admission and orientation materials; (l) intake screening records; (m) mental health intake screening records; and (n) medical intake screening record form. *Note:* extra mug shots, Social Security cards, driver's licenses, and similar documents may also be stored in this section, but these could also be filed in the records office because they will be sent to the Receiving and Discharge Department upon the inmate's release.
>
> *Section 4: Conduct, Work, Quarters Reports.* Contains: (a) disciplinary records (in chronological order); (b) incident reports and supporting materials (in chronological order by incident), including the incident report, inmate rights and responsibilities form, notice of disciplinary hearing form, duties of staff representative form, waiver of appearance form, disciplinary hearing report, disciplinary hearing checklist, administrative segregation order form, special housing unit record form, special housing review form, and temporary placement in disciplinary detention order form (including any disclosable supporting documentation); (c) work evaluation reports; (d) housing reports; (e) education data, including school progress notes, Adult Basic Education (ABE) scores, ABE certificate, General

continues

Exhibit 1-2 continued

> Equivalency Diploma (GED) test results, GED certificate, vocational training certificates; and (f) drug abuse program forms and correspondence.
>
> *Section 5: Release Processing.* Contains: (a) institution and unit release preparation checklists; (b) relocation request and approval; (c) request and approval of release plan from parole officer; (d) parole certificate and request form for same; (e) other release correspondence; (f) deportation documentation and correspondence; (g) mandatory release forms and certificate; (h) release authorization; (i) record of release funds disbursement; (j) law enforcement notification form; (k) release of immigration detainee with supervision to follow; (l) conditions of supervised release; (m) halfway house application, documentation, acceptance; (n) furlough application, approval, and other forms; (o) medical evaluation for halfway house placement; and (p) condition of halfway house placement form.
>
> *Section 6: General Correspondence.* Contains: (a) general and miscellaneous correspondence; (b) correspondence related to furloughs (other than documentation included in Section 5); (c) escorted trip documentation; and (d) inmate requests to staff.

summary; (g) medical, psychological, and psychiatric reports, if separate from classification material and disclosable; (h) judge's reports; (i) institution progress reports; (j) escape reports; (k) the front side of all incident reports, with disciplinary hearing packets given to inmate; (l) application for parole; (m) correspondence regarding release planning, aftercare arrangements, detainers, and halfway house transfers (including adjustment reports); (n) documents on fines; and (o) documents on other prerelease matters, including correspondence concerning parole consideration.
- Material exempt from disclosure to the inmate must not be placed in parole board files.

6. Judgment and Commitment Files

- The records office is responsible for maintaining a judgment and commitment file on each inmate, containing the following documents: (a) *originals* of the Judgment and Commitment Orders; (b) official correspondence with the courts or other regarding sentencing matters; (c) *original* parole orders; (d) negatives of all inmate photographs; (e) FBI and police records (rap sheets); (f) court documents concerning sentence appeals; (g) *original* release papers, including writs and warrants; and (h) *originals* of any escape notices.

7. Other Files

In addition to the central file, the parole file, and judgment and commitment file, the institution is likely to maintain other program-related files on individual inmates. The Health Services Department should maintain a medical file on each inmate, the Mental Health Department should maintain a mental health file, and the Education Department should maintain an educational and vocational training file. These files should be consolidated with other inmate files when the inmate is transferred to another facility or released.

8. Multiple Volumes

On occasion, an inmate's central file will become so voluminous that it will become necessary to create a second volume of the file. The second volume should be set up in the same six-part fashion as the first volume and should contain copies of all core documents required for an initial file. Core documents that should be placed into a second volume of an inmate's central file would include:

- the intact privacy folder from the first volume (i.e., a new privacy folder should not be created; the privacy folder should be transferred from the first volume to the second volume)
- the Special Monitoring System tracking card
- any sentence computations
- copies of the Judgment and Commitment Orders
- a chronological summary
- the presentence investigation report
- a chronological disciplinary record

The same procedures should be followed if additional volumes should be required.

Multiple volumes should be identified clearly on check-out cards, file tabs, and the file cover (i.e., "Volume 1 of 3," "Volume 2 of 3," "Volume 3 of 3," etc.).

9. Location and Storage

- Inmate files should be maintained at the current or last institution of confinement. After an inmate is released, the file should be

handled in accordance with agency-wide records management procedures. Typically, however, an inmate file is retained at the last institution of confinement for one year following an inmate's release; thereafter it is sent to the records office at agency headquarters or retired to an archival facility.
- Inmate files (whether they are central files held in the unit or judgment and commitment files held in the institution records office) should be arranged alphabetically by the last name of the inmate and stored in fireproof, locked cabinets that must be secured when members of the staff are not present. Inmates must never have access to those cabinets.
- At the warden's discretion, inmate central files in minimum-, low-, and medium-security facilities may be stored, in a secure fashion, within the individual units. It is recommended, however, that inmate central files in high- and maximum-security facilities be stored in a secure, centralized location (usually the records office). At a minimum, files must be located and secured behind a grille door.

10. Maintenance, Security, and Access Procedures

- The records office manager has overall responsibility for file retention and disposal, for certification of file contents, and for making files available to the courts upon request. Wherever inmate files are stored on a decentralized basis, supervisory authority is delegated to the unit managers, who are held accountable for file security, control, and maintenance. Files should be stored in secure cabinets, in secure areas, when not in use. Fireproof cabinets are recommended.
- Files must never be left unsecured or handled in such a way as to be accessible to any unauthorized persons. In particular, the staff must ensure that inmates are never permitted to transport or otherwise handle inmate files or any other confidential materials.
- All files must be returned to the filing cabinet prior to the close of business or tour of duty of each user. Files should never be left overnight in desks, offices, or other areas, even if those areas can be locked. This is of particular importance given the need for files during emergency situations. Files left in an individual staff member's desk, rather than returned to their proper storage place, may not be accessible or easily located if an emergency arises.
- A file check-out card should be prepared for each volume of each inmate file. Check-out cards should be approximately the same

height and length as the front of a volume, with a raised tab with the word "Out" printed on it. Check-out cards should be clipped to their respective volumes. Whenever a particular volume is borrowed, the user should sign and date the check-out card and leave the card in the volume's place. When the volume is returned to the filing cabinet, the user should cross out his or her name and reattach the check-out card to the volume. This system provides a tracking method when files are missing or needed for other purposes and improves the ability to take inventory.
- All inmate files must be counted whenever a filing cabinet is unlocked and must be recounted before the cabinet is relocked, if any significant period of time elapses between the unlocking and relocking. A file count record should be maintained that would note the date and time of the count, the total number of volumes counted, the initials of the counter, and the reasons for any changes since the most recent previous count. Because *each volume* must be counted separately, and because the files on many inmates will consist of two or more volumes, the total file count may exceed the total inmate count.

In addition, a name roster and census count must be conducted at least weekly. This count involves comparing a name roster of inmates against the files in each filing cabinet. Staff should place their initials beside each name on the roster that has a matching file, and the initialed roster should be retained for at least ninety days.

Files for inmates not housed in the institution (e.g., inmates out on writ, in home confinement, or in halfway houses) should be accounted for in the same manner. To facilitate accountability, files on these inmates should be kept in a separate section of the filing cabinet.

11. Uses of Inmate Files

Inmate files may be used for the following purposes:

- Officers and other employees of the correctional agency may use inmate files in order to obtain information in the performance of their official, assigned duties.
- Law enforcement officials or government attorneys may use inmate files as a source of information for investigations, criminal prosecutions, civil court cases, or regulatory proceedings.

- Information from inmate files may be publicly disclosed as a matter of routine *only* if it concerns facts in the public record (such as the inmate's name and offense, sentence data, location of confinement, and release date), and if the disclosure does not violate privacy statutes. Information may be made available to the news media on the same basis.
- Information from inmate files may be disclosed to contracting, consulting, or correctional agencies that have responsibility for confining, transporting, or otherwise securing inmates.
- Information from inmate files may be provided in response to *official* requests from legislators or legislative bodies (such as committees); ordinarily, requests would be considered official if they come from legislators or committees that have clear oversight responsibilities for corrections or the administration of justice. Information may also be provided to legislative staff members acting on behalf of a legislator or committee.
- Courts, court officials, and probation or parole officers acting in the official performance of their duties may be provided with information from inmate files.
- Victims or witnesses, pursuant to applicable victim/witness legislation, may be provided with information regarding an inmate's furloughs, appearances before the parole board, parole, halfway house transfers, expiration of sentence, escapes (including apprehensions), death, or other information related to release.
- Data specifically identified by statute may be provided to the Social Security Administration, the Veterans Administration, or to state agencies for the purpose of determining if an inmate is eligible for benefits under any programs managed by those agencies, with the understanding that data from the inmate file may not be retained by the requesting agency after the eligibility determination is made.
- Employees, former employees, and representatives of current or former employees of the correctional agency may obtain information from inmate files that pertains to any adverse or disciplinary personnel action being taken or considered against that current or former employee. To protect the inmate's privacy, however, that information may be sanitized, specific documents may be redacted, and appropriate protective orders may be requested to prevent further dissemination. Information from inmate files may also be provided, under the same conditions, to administrative agencies, arbitrators, or courts of competent jurisdiction

involved in any kind of action or lawsuit relating to adverse or disciplinary actions against a current or former employee.
- Employees of the archives, records administration, or hall of records in the state or county may have unrestricted access to inmate files in the performance of their statutory authority to oversee the creation, maintenance, and disposition of official government records.

All written or oral disclosures of information from an inmate file, apart from disclosures made to officers and other employees in the course of their official, assigned duties, should be documented on a form prepared for that purpose and that form should be placed in the inmate file.

Staff in a position to make decisions over whether or not to disclose information from inmate files should be familiar with all relevant freedom of information statutes and personal privacy statutes in their jurisdiction. They should carefully ensure that disclosures or nondisclosures are made in accordance with those statutes.

Line staff who have access to inmate files but whose official duties do not include making decisions regarding disclosure of information must be advised that personal information about inmates that they acquire in the course of their duties is to be regarded as confidential. Privacy statutes may carry criminal and civil penalties for the unauthorized release of information.

12. Review of Files by Inmates

- Inmates may obtain access to their own central files under two methods. First, the inmate or the inmate's representative may request access under the Freedom of Information Act, if one has been enacted in the particular jurisdiction. Such requests and determinations regarding disclosure are handled formally, under regulations of the correctional agency and in accordance with statutory requirements. Second, informal administrative procedures may be implemented that would grant inmate access to portions of their file, in accordance with sound correctional practices and concerns and at the warden's discretion.
- To facilitate inmate access to information under informal administrative procedures, the warden should designate staff who would be responsible for reviewing inmate files before they are made available to inmates, for receiving inmate requests to review files, and for monitoring inmates while they are reviewing files.

- An inmate seeking to look at his or her file should submit a request to the staff member designated under local regulations. The designated staff member should provide written acknowledgment of all requests, and the inmate should be permitted to review the file whenever practicable.
- All nondisclosable materials must be removed from the file before the inmate is permitted to review it. This would include the privacy folder in its entirety. In addition, parole files are not disclosable, except as specifically authorized by the parole board.
- All file reviews by inmates must be done under constant and direct staff supervision. An entry should be made on the inmate activity record portion of the file to show the date that the inmate reviews the file. The staff member who monitors the review should initial the entry and should request the inmate to initial it as well.
- Inmates may not obtain copies of documents in their files under these informal administrative procedures. Instead, inmates desiring copies must submit formal requests under Freedom of Information statutes.
- Inmates may challenge the accuracy of information in their central files. Unit staff should take reasonable steps to ensure the accuracy of challenged information, particularly in cases where this information can be verified. Initial reasonable steps include requiring specific information from the inmates, such as providing documents that support a challenge or the names of individuals who can shed light on challenged information. If the inmate is able to support a challenge, the unit staff must investigate the matter and take steps to correct inaccuracies. If the inmate challenges information in a document that was not created by the institution or the correctional agency (such as a court or a parole board document), then the staff should inform the office that originated the document of the inmate's challenge. If that office confirms that the document is inaccurate, staff should insert a memorandum to that effect in the file to ensure that no decisions affecting the inmate are based on the discredited information.

13. Transfer of Records between Institutions within the Correctional Agency

- Files on inmates who are transferred to other institutions within the correctional agency, or are returned to the custody of the correctional agency as parole violators, mandatory release violators, or escapees, and are designated to other institutions, should

be consolidated and shipped to the new institution immediately. Files will be shipped by certified mail, with a return receipt requested. Requests for the transfer of files should be documented, processed, and receipted in the same manner as requests for transfers of inmates. Telephonic requests should only be honored if the identify of the requesting party can be verified.
- When an institution receives inmate files that are transferred from another facility, the judgment and commitment files should be sent to the records office, the medical records should be sent to the hospital administrator, the mental health records should be sent to the chief of mental health services, the educational and vocational training records should be sent to the Education Department, and the central files and parole files should be sent to the appropriate unit (depending on the facility's security level).

14. Retirement of Inactive Files

- As noted above, the parole file should be shipped to the appropriate parole board office having supervisory authority over an inmate, following that inmate's release.
- Inmate central files should remain in the unit for approximately two weeks following an inmate's final release, to give sufficient time for final release papers to be consolidated and filed. Two weeks after an inmate's final release, the central file, judgment and commitment file, education file, medical file, and mental health file should be sent to the records office and consolidated. The outside of the file folder should be stamped with the year of the expiration of the sentence. The files should remain in the custody of the records office pending disposition in accordance with regulations and statutes for handling official government records. The government archives, records management agency, or hall of records should be consulted for specific guidelines. Typically, however, an inmate's file will be housed by the institution's records office for one year after the inmate's release, at which point they should be transferred to the correctional agency's headquarters or to an archival facility or hall of records. Depending on guidelines adopted by the state or county government, those files may be permanently retained or destroyed after a given period of time (usually twenty-five to thirty years, at a minimum, following the expiration of the sentence).

CHAPTER 2

Day-to-Day Supervision and Security Procedures

I. INMATE ACCOUNTABILITY

Overview

Inmate accountability should be maintained through the use of easily retrieved identification materials on each inmate, through a series of multiple scheduled head counts throughout the day, and through unscheduled head counts. More frequent head counts should be conducted for special management cases, and certain inmates—such as those considered suicidal—should be under continual observation. Procedures should also be implemented to ensure accountability for inmates assigned to work release, educational release, furloughs, and other temporary absences.

1. Control Center Records

Picture cards for all inmates assigned to an institution should be on file in the control center, for quick reference. Control center records should note housing assignments, job assignments, custody, sentence data, and other necessary security and control information.

2. Head Counts

The count system should provide for at least five or six official counts during each twenty-four-hour period. All staff conducting counts must personally observe human flesh for each inmate counted to ensure that a dummy is not being counted. All counts should be double-counted—that is, two staff members should conduct counts and both should sign the count slip. All count slips are to be written in ink and forwarded to the control center. Count sheets should be consolidated into a master count sheet, prepared in ink by the control center officer.

3. Lock-Down Accountability Checks

Institution-wide lock-down accountability checks should be conducted at least once a month to determine if inmates are in authorized areas (work areas, housing units, classrooms, recreation yards, medical facilities, food service areas, etc.). The accountability check is to be announced on the public address system at a random date and time, with no prior notification. As soon as the announcement has been made, staff must secure all entrances and exits. No inmate movement should be permitted following the announcement. Staff should then survey their areas of supervisory responsibility in order to determine or verify: (a) the total number of inmates present, (b) the total number of inmates authorized, and (c) the total number of inmates unauthorized, with their names and registration numbers.

Although the lock-down accountability check is not designed to be a formal head count, every effort should be made to arrive at a total institution head count. If this is not accomplished within a reasonable time, the warden may resume normal operations, with the understanding that an accountability problem exists and needs to be rectified.

The shift commander is responsible for supervising the taking of the accountability check and documenting the results (such as time elapsed, discrepancies noted, and actions taken) in the shift commander's log.

4. Detail Accountability Checks

Detail accountability checks (or area checks) are similar to lock-down accountability checks but are conducted only for specific areas of an institution. They should be carried out for at least fifteen inmate assignment locations every week. Institutions with fewer than thirty inmate assignment locations should conduct checks in at least 50 percent of them every week.

Area checks should be limited to one area at a time (such as the paint shop, a single housing unit, the hospital, the kitchen, or other areas that can be isolated) and should not interfere with overall institution operations. They should be staggered so as to avoid a predictable pattern.

5. Daily Change and Transfer Sheet

A list showing all changes in inmates' status, including changes in housing units and transfers, should be compiled in the control center and distributed daily to all areas of the institution.

6. Inmate Call-Outs

The daily call-out sheet lists any inmate who has been instructed by a staff member to report to a specific area. Information for the call-out sheet should be obtained from the institution's automated data system. The call-out sheets should be distributed to all concerned staff and posted in the housing units for the inmates. Special precautions must be observed to ensure that inmates are not able to circumvent procedures or make additions or deletions to the posted call-out sheets that would undermine the accountability system.

The staff member placing an inmate on call must ensure that the inmate arrives at the specified area on time. If the inmate does not arrive as scheduled, the staff member who requested the call-out should contact the staff member responsible for the inmate at that time. If an inmate does not report as instructed, he or she should be reported to the shift commander as missing, and immediate action should be taken to locate the missing inmate.

7. Pass System

A well-maintained pass system is an asset to any institution's accountability program. Institutions with a pass system should account for every pass on a daily basis. The pass book must be turned in daily, along with all the passes issued for that day. A staff member must check the used passes against the stubs or noncarbonized copies remaining in the book. A log is to be maintained in the control center to help identify repeated discrepancies in the pass system. All pass system discrepancies are to be reported to the shift commander, who will interview any inmates or staff who fail to comply with procedures and will note in the log any action taken. Frequent discrepancies should be brought to the attention of the chief of security, who will notify the relevant department head.

8. Front and Rear Entrance Pass

Any inmate with an assignment that requires passing through the front or rear entrance must have an approved gate pass card, which will bear the inmate's photograph, custody classification, work assignment, offense, and sentence. The gate pass should be signed by the chief of security and the deputy warden responsible for security. Each card should be stamped with an imprint seal and laminated for durability. The imprint should cover a portion of the inmate's photograph, to prevent tampering with the gate pass. Blank gate passes must be accounted for in the strictest fashion and secured in a locked safe or a locked filing cabinet. Procedures for destroying outdated gate passes shall be established and closely observed.

9. Detail or Work Crew Kit Cards

Cards on each inmate assigned to a work crew should be maintained in the control center. A crew kit card should have a photograph of the inmate and include his or her name, registration number, job assignment, housing assignment, custody level, medical status, and any other special conditions. Inmates must never be permitted to handle the crew kit cards.

10. Special Accountability

Any inmate confined in continual locked status, such as administrative segregation or disciplinary detention, should be observed by a staff member at least once every thirty minutes (on an irregular or unpredictable schedule). Closer observation is required for each inmate who is considered suicidal, who is mentally ill, or who demonstrates unusual or bizarre behavior.

II. SEARCHES

Overview

It is necessary to conduct searches of inmates, inmate housing areas, and inmate work areas in order to locate contraband and to deter the introduction and movement of contraband. When conducting searches of an inmate's person, staff must avoid unnecessary force and strive to preserve the dignity of the individual being searched. Staff should employ the least intrusive method of search practicable, as necessitated by the type of contraband and the method of suspected introduction.

1. Body Searches of Inmates

- A *pat search* is an inspection, using the hands, of an inmate's person and his or her clothing and personal effects that does not require the inmate to remove clothing. The staff may conduct pat searches of inmates on a routine or random basis in order to control contraband.
- Metal detection devices may be installed in the institution to assist in efforts to control contraband. Metal detector searches of inmates' persons may be conducted under the same circumstances as a pat search or in addition to a pat search.
- A *visual search* or *strip search* is a visual inspection of all body surfaces and body cavities. Staff may conduct a visual search whenever there is a reasonable belief that contraband may be concealed on the person or that a significant opportunity for concealment has occurred. For example, placement in a special housing unit, leaving the institution, or reentering an institution after possible contact with the public (a community trip, court transfer, contact visit in an institution visiting room, etc.) is sufficient to justify a visual search. The visual search should be conducted in a manner designed to preserve as much of the inmate's privacy as practicable. Except in minimum security institutions, inmates must undergo a visual search when leaving the institution for any reason (even when being released). Visual searches should also be conducted on inmates when they are being processed into the institution through receiving and discharge, when inmates are being placed in special housing or the

Special Management Unit, and when inmates are returning from an outside work detail.
- Ordinarily, visual searches should only be conducted by a staff member of the same gender as the inmate being searched. A staff member of the opposite gender may conduct a visual search, however, if a staff member of the same gender is unavailable to conduct the search *and* if circumstances are such that a delay in conducting the search would risk loss of the contraband. In cases where a staff member of the opposite gender conducts a visual search, the incident must be documented in the inmate's central file and the reasons for the opposite gender search much be clearly stated.
- A *digital search* or *simple instrument search* is an inspection of any body cavity, using fingers or simple instruments—such as an otoscope, tongue depressor, short nasal speculum, or simple forceps—to probe for contraband or any other foreign item. A digital or simple instrument search may be conducted only by designated and qualified health care personnel (for example, physicians, physician assistants, and nurses), only upon the approval of the warden or acting warden, and only if there is reasonable belief that an inmate is concealing contraband in or on his or her person. All digital and simple instrument searches, and the reasons for those searches, must be documented in the inmate's central file.
- If located in a body cavity, the contraband or foreign item may be removed immediately by medical staff, if such removal can be performed using fingers or simple instruments. Staff may not conduct a digital or simple instrument search if it is likely to cause physical injury to the inmate.
- Persons of the opposite gender from the inmate, except for health care personnel, may not conduct or observe a digital or simple instrument search.
- Staff should solicit the inmate's written consent prior to conducting a digital or simple instrument search, but the inmate's consent is not required.

2. Close Observation (Dry Cell Status)

- When reasonable belief exists that an inmate has ingested contraband or concealed contraband in a body cavity, and a digital search or simple instrument search are inappropriate or likely to cause physical injury to the inmate, the warden or designee may

authorize the placement of the inmate in a room or cell to enable staff to observe the inmate closely, until such time as the inmate has voided the contraband or until sufficient time has elapsed to preclude the possibility that the inmate is concealing contraband. Such placement is commonly referred to as *dry cell status*.
- The warden or acting warden may authorize dry cell status, during regular duty hours or at other times, if one or the other is present in the institution. At other times—for example, during nonduty hours—the shift commander may authorize dry cell status, in consultation with the duty officer.
- The length of dry cell status should be determined on an individual basis. Ordinarily, the chief of security—in consultation with qualified medical personnel—should decide when it is appropriate to terminate dry cell status. Automatic review of dry cell status by the segregation review officer should occur, however, when an inmate has been in dry cell status for more than three days, and by the warden when an inmate has been in dry cell status for more than seven days. The chief of security and qualified medical personnel should be consulted in the course of these reviews.
- Dry cells should meet the following requirements: (a) they must be free of hiding places and equipped with no furniture other than a bed; (b) doors to the cell must have observation panels to protect staff and permit unobstructed observation; (c) windows should have a security screen to prevent the disposal of contraband; and (d) cells equipped with a toilet and/or sink must have the water shut off before the inmate is placed in the cell.
- A dry cell must be completely searched before an inmate is placed there, to verify that it is free of contraband. Potential hiding places for contraband in the cell, if any, should be noted and given special attention during subsequent searches.
- Staff members assigned to supervise an inmate in dry cell status must be of the same gender as the inmate and must maintain complete and constant visual supervision of the inmate. Supervising staff members must ensure that the inmate is never allowed freedom to move around unsupervised or given the opportunity to dispose of any objects he or she may be concealing. Supervising staff members should be issued a portable radio and a flashlight, should make watch calls by radio, and should maintain a daily log and a special housing unit record on the inmate. The shift commander is responsible for ensuring that supervising staff review dry cell status post orders prior to assuming the post and for

providing a relief person so that the inmate is not left unattended during the relief period.
- The staff member responsible for initiating the dry cell watch should advise the inmate of the conditions and of what is expected. This notification should be documented on a form developed for that purpose and placed in the inmate's central file.
- The inmate should be given a visual search before being placed in the dry cell and issued a jumpsuit or other suitable loose-fitting clothing. At least once each shift the inmate should be given a visual search and the dry cell itself must be thoroughly searched. The staff should take care not to set a pattern. Before conducting such searches, the shift commander must be notified and the relief person sent in to provide inmate supervision during the search.
- The inmate must provide a urine sample within two hours of placement in dry cell status and must provide a second urine sample before being released from dry cell status.
- The inmate may not be allowed to have personal property while in dry cell status, apart from legal and personal mail and a reasonable amount of legal materials, when so requested. Personal hygiene items should be controlled by the staff.
- No inmate in dry cell status may be allowed to come into contact with another inmate. Neither visiting nor nonemergency telephone calls are permitted. Inmates in dry cell status may not be permitted recreation outside the cell.
- The dry cell should be illuminated at all times.
- Trash may not be allowed to accumulate in the dry cell, and each item being discarded should be carefully searched before disposal.
- When the inmate is in bed, he or she must lie on top of the mattress in full view, room temperature permitting. If it is necessary for the inmate to be covered with a blanket or sheet, then his or her hands must remain outside the blanket or sheet and visible at all times, so that staff may observe any attempt to move contraband. An inmate might attempt to remove contraband from a body cavity or to insert contraband into a body cavity, so supervising staff must keep the inmate's hands under continual observation.
- The inmate should be served the same meals as those available to the institution's general population. All meals should be inspected for contraband before being delivered to the inmate. Any food remaining after the meal, as well as the utensils and tray,

must be thoroughly inspected before being returned to food service.
- The inmate may receive no medications, apart from those prescribed and administered by institution medical personnel. No laxatives may be given apart from natural laxatives, such as coffee, prune juice, etc., that would be part of the inmate's normal meals.
- Institution staff must be available to the inmate upon request, within reason and within the bounds of security concerns. When the inmate requests to shave, brush teeth, etc., a wash pan and a container of water should be provided for use in the cell.
- When the inmate needs to urinate or defecate, he or she should be furnished with a clean, empty hospital bed pan. Staff supervising the inmate should notify the shift commander, who will direct a second staff member to provide supervision. Using rubber gloves, and a forceps or tongue depressor, the staff should inspect the excretion closely to ascertain whether any contraband is present.
- The staff must notify the shift commander when contraband is found and follow procedures outlined in "Discovery of Contraband," below, for handling contraband.

3. X-Ray, Major Instrument, Fluoroscope, and Surgical Intrusion Searches

- The institution physician may authorize use of a fluoroscope, major instrument (such as an anoscope or vaginal speculum), or surgical intrusion for medical reasons only, and only with the inmate's consent.
- When no reasonable alternative exists, and an x-ray examination is deemed necessary for the security, good order, or discipline of the institution, the warden—with the approval of the appropriate executive-level official of the correctional agency—may authorize the institution physician to order a nonrepetitive x-ray examination for the purpose of determining if contraband is concealed in or on the inmate's person (for example, in a cast or a body cavity). The x-ray examination should not be performed if the institution physician determines that it is likely to cause serious or lasting medical harm to the inmate. Staff should solicit the inmate's consent prior to the x-ray examination, but such consent is not required under these specific circumstances.

- The warden may direct x-rays of inanimate objects in cases where the inmate is not exposed.

4. Searches of Inmate Housing Units and Work Areas

- Inspections of housing units and work areas are designed to detect contraband, prevent escapes, maintain standards of sanitation, and eliminate fire and safety hazards. Therefore, housing units and work areas should be searched routinely. These searches should be conducted on an irregular schedule so that inmates cannot anticipate when they will occur.
- Staff members may search an inmate's living area or work station, and the personal possessions found in those places, without notifying the inmate, without obtaining prior approval from the inmate, and without the inmate's presence. Written documentation of each housing unit search should be maintained within the individual housing unit.
- Work areas, including prison industrial facilities, should be searched each working day by shop supervisors, and those inspections should be supplemented with periodic searches by regular search teams. Documentation of daily searches is not required but should be informally monitored to ensure compliance. The chief of security, however, should maintain documentation of search team inspections.
- To the greatest extent possible, staff conducting searches should leave the areas they have searched—whether they are housing units or work areas—in their original order and condition.

5. Discovery of Contraband

- If any of the types of searches described above uncover contraband, the contraband must be secured in a plastic evidence bag and proper documentation and chain of evidence procedures must be maintained. The institution should have procedures in place to coordinate with law enforcement officials and prosecutors if the matter warrants criminal prosecution.

III. POSTED PICTURE FILES

Overview

Maintaining a posted picture file is a method of specifically identifying those inmates who pose a significant threat to inmate safety, staff safety, the security of the institution, or the welfare of the surrounding community because of their prior record, current offense, institutional adjustment, or other factors. Having images of these inmates readily available in the posted picture file alerts staff as to which inmates fall into these categories.

1. Criteria

- Posted picture files status is appropriate for any inmate who meets one or more of the following criteria:
  — Escape risk; has attempted escape from a secure institution within the previous three years or successfully carried out an escape from a secure institution within the past ten years
  — History of assaultive behavior; has been involved with a serious assault with a weapon within the previous three years, a physical assault in which grievous bodily harm was inflicted, or an assault or attempted assault upon a staff member
  — Homicide (within an institution)
  — History of sexual offenses; has committed a sexual assault (or attempted sexual assault) against another person in an institution within the previous five years or has a history of sexual advances toward anyone in an institution
  — Confinement in the Special Management Unit within the previous five years
  — Confirmed member of a disruptive group
  — Involvement in an incident related to the introduction or attempted introduction of drugs or hazardous contraband into an institution within the previous five years
  — Special skills or experience as a locksmith, gunsmith, explosives expert, security electronics technician, computer expert, etc., whether those skills or experience were acquired for legitimate purposes or through illicit activities

Inmates who have a history of manipulating or compromising staff, have led food strikes or work stoppages, are former judicial or law enforcement officials, or who otherwise might appear to pose a risk to institutional stability but do not fall into any of the categories cited above, may also be placed on posted picture status, provided the staff furnishes a written justification for such placement to the warden and obtains the warden's approval.

2. Responsibilities

- The deputy warden or associate warden in charge of security operations normally would be responsible for approving an inmate's placement in posted picture file status (except for inmates who do not fall under the major categories cited above, in which case the warden would be the approving official).
- The posted picture file is maintained by the chief of security and is kept in the shift commander's office for staff review. A duplicate copy of the posted picture file should be maintained in the investigative supervisor's office.
- The chief of security should institute procedures to keep the posted picture file current. Forms should be prepared to document an inmate's placement in posted picture file status and entries should be made to document the inmate's work and housing assignments. Forms should be removed from the posted picture file as inmates are transferred, released, or taken off posted picture file status.
- The institution's investigative section should maintain a database of all inmates in posted picture file status. This database should be software-compatible with similar databases in all other institutions within the system and in the correctional agency's headquarters. Because inmates in posted picture file status will often qualify for placement in the investigative section's tracking system, the investigative section should coordinate with the chief of security to ensure that its tracking system is consistent with the posted picture file system.

3. Reviews

- All security and unit staff should review the posted picture file at least once a month in order to be fully acquainted with which inmates are on that status. All other institution staff should review the posted picture file at least once per quarter. A bound ledger should be maintained in the shift commander's office; staff members should sign and date the ledger to verify that they have reviewed the posted picture file as required.
- The deputy warden or associate warden in charge of security operations, the chief of security, and the investigative supervisor should review the posted picture file quarterly to determine each inmate's suitability for continued posted picture file status. Each

review should be documented on the reverse of each posted picture file form.

IV. ALCOHOL TESTING

Overview

Correctional institutions should have surveillance programs in place in order to deter and detect the illegal introduction or use of alcohol. Alcohol surveillance programs should include procedures for monitoring and testing individual inmates or groups of inmates who are known or suspected users of alcohol, who are considered high risks based on observed behavior, or who may have had access to alcohol while outside the institution. Testing of inmates on a random basis may also be performed.

1. Procedures for Testing Inmates

- Alcohol monitoring programs should be coordinated and supervised by a senior management official of the institution, such as the chief of security. The appropriate level of monitoring and testing to deter and detect alcohol use will vary by institution. Alcohol surveillance needs are determined by each institution's security level and openness. Other important factors are whether the institution permits inmates to have any kind of community contacts that might provide opportunities for inmates to acquire alcohol, or if there are work areas in the institution where inmates may have access to ingredients that could be used in the manufacture of alcoholic beverages.
- Alcohol surveillance procedures should be implemented with particular care in work areas and housing areas where there are inmates who have histories of alcohol abuse. At minimum security institutions, daily testing on a random basis should be conducted on inmates returning from outside work details.
- A log of alcohol tests should be maintained in a bound ledger. The log should indicate the name and registration number of each inmate subjected to an alcohol test, the name of the staff member administering the test, the time and date of the test, and the test results. If the inmate refuses to cooperate in an alcohol surveillance test, the lack of cooperation should be noted in a separate

column in the log. The staff should submit an incident report on any inmate who shows a positive substantiated test result for alcohol or who refuses to cooperate in an alcohol test.
- Ordinarily, a 0.05 or higher blood-alcohol content reading should be considered positive for alcohol use. If an initial test provides a positive reading, then a second test should be administered fifteen minutes later. The inmate should be segregated for the fifteen-minute waiting period and must not be permitted to eat, drink, or smoke during that time. If the second test confirms a reading of 0.05 or higher, the staff member should prepare an incident report documenting the results.

2. Alcohol Testing of Suspect Liquids

A test reading of 0.05 should be considered positive for alcohol content. Equipment should be available for testing liquids suspected of containing alcohol. The corrections agency should designate specific brands or models of equipment that are approved in all institutions in the system for testing inmates and suspect liquids for alcohol. Staff using such testing instruments must be familiar with how to operate them, based on manufacturer's instructions. Calibration checks should be made on the testing equipment at least once a month, according to procedures outlined in the manufacturer's instructions. Calibration checks should be noted in the alcohol test log.

V. URINE SURVEILLANCE

Overview

Urine testing as a means to detect drug use should be administered to all inmates, on a random basis. In addition, regular urine testing should be conducted to monitor specific groups or individual inmates considered high risks for drug use, such as those involved in community activities, those with a history of drug use, and those who are directly suspected of using drugs. It is important that all aspects of urine testing, including urine collection, recording, mailing, and processing, be carried out by institution staff and not delegated to inmate clerks.

1. Test Categories

- Different categories of inmates should be identified for urine testing. The frequency of testing should depend on the risk of drug use presented by those categories. By targeting different categories, the institution should be able to assess potential problem areas and take appropriate action to deter unauthorized drug use. In particular, the urine surveillance program should identify specific inmates or groups of inmates suspected of using drugs and test them systematically over an extended period of time. A list of categories of inmates for testing is shown in Exhibit 2-1.
- Specific procedures need to be followed in selecting and testing inmates for the random group. A nonalphabetical list of randomly selected inmates, representing 5 percent of the total inmate population, should be generated every month by the institution's automated data system. All inmates are eligible for inclusion in the random list, even if they are already required to provide urine samples as part of another group being monitored. The coordinator of the urine surveillance program (normally the chief of security) should obtain the random list from the automated data

Exhibit 2–1 Categories of Inmates for Urine Testing

Inmates involved in community activities:	TEST 50%
Inmates with history of drug use, who are involved in community activities:	TEST 100%
Members of disruptive groups:	TEST MONTHLY FOR DRUGS AND MARIJUANA
Inmates who test positive for unauthorized drug use:	TEST MONTHLY FOR ONE YEAR FOLLOWING ANY INCIDENT OF UNAUTHORIZED DRUG USE*
Entire inmate population:	TEST 5% EVERY MONTH, ON RANDOM BASIS**

*An inmate should be removed from the suspect list if no further positive tests occur during the follow-up period. If another positive test occurs, then the inmate should remain on the suspect list for an additional twelve months from the date of that test.

**Inmates may be selected for random testing regardless of their involvement in a specific urine surveillance program.

system on the first working day of every month and should ensure that the list is properly secured.

The random list must be used sequentially, beginning with the first name on the list. No inmate whose name appears on the list may be excused from participating, if the inmate is capable of doing so. If even one inmate is excused, then the selection process becomes arbitrary instead of random.

To prevent the establishment of testing patterns that inmates can predict, there should be no regular schedule for collecting specimens. Samples should be taken at different hours of the day, and random lists can overlap from month to month. Staff may take up to six weeks to test inmates whose names appear on a random list for any given month. Therefore, it is possible that inmates whose names appear on the random list for one month may not all be tested until the middle of the following month—by which time testing has already begun on a new group of randomly selected inmates.

2. Sampling Procedures

- As noted above, regular schedules for collecting specimens should be avoided, so that inmates cannot predict and thereby compromise urine surveillance activities. Urine collection should take place in only one or two centralized areas in the institution, such as the shift commander's office or the Receiving and Discharge Department, by staff who are thoroughly familiar with specimen collection procedures. Inmates undergoing testing must provide a quantity of urine sufficient for performing the required laboratory tests (i.e., approximately a full specimen bottle—60 cc. or 2 oz.).
- When an inmate reports for testing, the staff must: (a) keep the inmate under direct visual supervision at all times; (b) search the inmate completely, in order to detect any device designed to provide a substitute for the requested sample or any substance that might possibly be used to contaminate the sample; (c) require the inmate to wash his or her hands before providing the sample; (d) make a positive identification of the inmate; (e) collect the sample from the inmate; (f) assign a urine sample identification number to the specimen; (g) label the urine bottle with that identification number and with the date; (h) record the urine sample identification number next to the inmate's name on the

laboratory slip; (i) request the inmate to verify the urine sample identification numbers on the bottle and on the laboratory slip and initial both (if the inmate refuses to initial these items, then a second staff member should verify the numbers and initial the bottle and the slip); (j) keep the specimen under direct observation until it is moved to a locked area where it may be stored securely until being mailed to the laboratory.

- A staff member of the same sex as the inmate should directly supervise the giving of the urine sample. If an inmate is unwilling to provide a urine sample within two hours of being instructed to do so, the staff should file an incident report. The chief of security may be authorized to extend this two-hour period if extenuating circumstances exist (for example, if an inmate is dehydrated or is suffering from a documented medical or psychological problem). No extra time need be allowed, however, for an inmate who directly and specifically states his or her refusal to provide a urine sample.
- To assist the inmate in providing a urine sample, staff should offer the inmate eight ounces of water to drink at the beginning of the two-hour time period. An inmate may be given an additional eight ounces of water to drink during the two-hour time period if he or she requests it.
- To eliminate the possibility of urine samples being diluted or adulterated, staff members should keep the inmate under direct visual supervision during the entire two-hour period or until a complete sample has been furnished. Where practicable, staff members may consider indirect rather than direct supervision of an inmate who claims to be willing to provide a urine sample but is unable to do so because of being under direct visual supervision. The inmate (clothed in underwear only) should be thoroughly searched and then placed in a "dry room" (i.e., a room or cell with no plumbing or water) to provide the sample.
- An inmate is presumed to be unwilling to provide a urine sample if he or she fails to do so within the allotted time period. The inmate should be given the opportunity to refute this presumption during the disciplinary process.
- If a positive test occurs, the staff should determine whether there is a justifiable or acceptable reason, such as the inmate's use of prescribed medicine. If the inmate's urine test shows a positive result for the presence of drugs for which no justifiable or acceptable reason can be found, then the staff should submit an incident report. This incident report merely provides written documenta-

tion of a positive urine test result that cannot be justified. A second test to confirm a positive reading must be conducted before an incident report may be filed charging the use of a particular drug, and the second test may be taken only after a minimum waiting period has elapsed (see "Screening and Confirmation," below). Photocopies of the laboratory report and the laboratory slip completed at the time the specimen was collected, listing the inmate's name and urine sample identification number, should be attached to the incident report.
- Retesting at the inmate's request is not permitted on a positive result.

3. Screening and Confirmation

- Analysis of urine samples should be carried out, on a contract basis, exclusively by the laboratory approved by the correctional agency for performing and reporting urinalysis. Urine samples must be mailed to the testing laboratory no later than the first working day after they are collected.
- The urinalysis laboratory should screen for the following drugs and metabolics at the minimum levels established in the contract: (a) morphine (total, free, or glucuronide); (b) methadone and metabolite; (c) codeine; (d) other opiates; (e) barbiturates (including, but not limited to, amobarbital, phenobarbital, pentobarbital, butabarbital, hexobarbital, and secobarbital); (f) amphetamines (including, but not limited to, d-amphetamine and methamphetamine); (g) cocaine (free); (h) cocaine metabolite (benzoylecgonine); (i) phencyclidine; and (j) marijuana.
- Positive results must be confirmed with follow-up urine tests before disciplinary action may be taken. A waiting period must elapse, however, between the initial positive test and the follow-up confirmation test. The waiting periods vary by the type of drug and correspond to the maximum lengths of time that particular drugs would appear in a person's urine after the last use. A list of waiting periods for retesting is found in Exhibit 2–2.

4. Reporting Procedures

- As coordinator of the urine surveillance program, the chief of security should be responsible for providing a monthly report on urine surveillance activities (covering the preceding month) to

Exhibit 2-2 Waiting Periods between Initial Positive Test and Follow-Up Confirmation Test, by Drug Type

Amphetamine, methamphetamine, cocaine, cocaine metabolite:	3 DAYS
Methadone, methadone metabolite:	5 DAYS
Morphine, codeine, opiates, meperidine (Demerol), pentazocine (Talwin), propoxyphene (Darvon):	6 DAYS
Barbiturates, phencyclidine (PCP):	11 DAYS
Phenobarbital:	14 DAYS
Marijuana (THC):	30 DAYS

the appropriate senior-level official at the correctional agency's headquarters. That official at headquarters, in turn, should prepare a monthly report showing data for each institution and aggregate data for the entire system.
- The monthly report should provide statistics in the following categories: (a) number of urine samples taken; (b) number of positive test results; (c) number of inmates detected for unauthorized drug use; (d) positive test results, by type of drug; and (e) number of refusals to provide urine specimens.
- Statistics in each of these categories must be broken out by the various sampling groups (inmates involved in community activities, inmates with a history of drug use, members of disruptive groups, inmates who test positive for unauthorized drug use, and the entire inmate population). In addition, the statistics from all categories should be tallied in order to provide a total figure for the entire institution.

VI. THE SPECIAL MONITORING SYSTEM

Overview

To ensure institutional security and public safety, the movement and activities of certain inmates who present exceptional management problems need to be monitored by specially designated staff at the correctional agency's headquarters. These inmates are known as *special monitoring cases*. Headquarters clearance is required for any transfers, temporary releases (such as releases on writs), or community activities involving special monitoring cases. The purpose of the special monitoring system is to coordinate special case management actions and to provide consistent application of agency policy in cases where there are particularly unusual or extreme circumstances that must be considered.

1. Responsibility

- The appropriate executive-level official at the correctional agency's headquarters (such as a deputy commissioner or deputy director in charge of operations) should designate a special monitoring coordinator at headquarters, with responsibility for overseeing all special monitoring decisions and activities. The wardens of each facility in the system should designate a special monitoring coordinator for inmates confined in their respective facilities. Community program managers should be designated as special monitoring coordinators for inmates confined at contract facilities.

2. Assignments and Classification Procedures

Special monitoring cases are classified according to the assignments listed in Exhibit 2–3. An inmate may be classified as a special monitoring case by a community program manager or special monitoring coordinator at any time, provided the classification is substantiated by factual information. Appropriate sources of factual information would include presentence investigations; police reports; reports on convicted offenders provided by sentencing judges; documentation on an inmate's institutional behavior from his or her current central file or from a central file from a previous incarceration; newspaper clippings substantiating an inmate's notoriety; or from communications from the U.S. Justice Department's Office of Enforcement Operations for special witness cases, correctional agencies in other states, or the U.S. Secret Service. An inmate's self-admission, body tattoos, etc., may be considered when classifying an inmate as a member of a security threat group. Ordinarily, charges for which an inmate has not been found guilty may not be considered in the special monitoring classification process.
- An inmate's classification as a special monitoring case becomes effective as soon as the classification (including the appropriate assignment category) is entered into the automated inmate infor-

Exhibit 2–3 Special Monitoring Cases As Classified by Assignments

Special Witness Cases. These are inmates who have agreed to cooperate with law enforcement, judicial, or correctional authorities. As witnesses or potential witnesses against persons or groups involved in illegal activities, their lives or safety may be in jeopardy. Inmates may be granted special witness status by the prosecutor's office (under appropriate state statutes) or by the correctional agency's executive official, who is responsible for the special monitoring program. Participation in a special witness program is completely voluntary on the part of the inmates.

Sophisticated Criminal Activity. Inmates in this category have been involved in sophisticated, large-scale criminal activities. Examples of sophisticated criminal activities include major drug offenses, street or prison crimes, property offenses, and white-collar offenses. Classification in this category requires that offenders were the principal figures or prime motivators in the criminal organization or activity from which substantial income or resources could have been obtained. Benchmarks for determining if income or resources realized through criminal ventures should be considered "substantial" would be $5,000,000 or more for drug offenses and $1,000,000 or more for property offenses or white-collar crimes.

Threats to Government Officials. This category covers inmates that the U.S. Secret Service, the state police, or local police have identified—in writing—as requiring special surveillance because of threats they have made to government officials.

Unusual Notoriety. Inmates in this category have received exceptionally widespread publicity (such as national media coverage) because of their criminal activity or because they were public figures.

Security Threat Groups. Inmates in this category belong to or are closely affiliated with groups, such as prison gangs, that have a history of disrupting operations and security in either federal or state correctional facilities.

Federal and Out-of-State Prisoners. Inmates in this category have been accepted into the custody of the correctional agency for the service of sentences imposed in federal courts or sentences imposed in other states. This includes inmates who are cooperating witnesses for other states, as well federal and state offenders being boarded under regular contracts.

Separation Cases. Inmates in this category must not be confined in the same facility with other specified inmates or potential inmates. Inmates who may be assigned to this category would include those who have given court testimony against a specific inmate or have otherwise provided information to law enforcement authorities about a specific inmate's involvement in illegal activities, as well as inmates who have been the subject of testimony given by other inmates. This assignment may also include inmates who are to be separated from other specific individuals at the request of the courts, prosecutors, or the police.

Special Supervision Cases. Inmates in this category require special management attention but do not warrant assignment in any of the other categories. This may include inmates who have worked in law enforcement, have been involved in a hostage situation, or have belonged to terrorist groups.

mation system. The case manager must provide the inmate promptly with written notification of classification as a special monitoring case and the reasons for the classification. Special witness cases

should be notified during the commitment interview. The inmate should be requested to sign a form verifying that he or she has received notification of special monitoring status; if the inmate refuses to sign the notification form, then staff witnessing the refusal should indicate this fact on the form and sign it.
- Staff may limit the amount of information given to an inmate regarding the reason for a special monitoring classification if it is in the best interests of safety or security. For example, if an inmate has been identified as a separation case to prevent that inmate from coming into contact with another inmate who has provided testimony or otherwise informed against him or her, then the name of the inmate informant should not be given to the affected inmate as part of the explanation for classification as a special monitoring case.
- If the classification of an inmate as a special monitoring case is made at the headquarters level, then further review is unnecessary. Headquarters must notify the appropriate institution immediately when it classifies an inmate designated to that institution as a special monitoring case.
- If classification of an inmate as a special monitoring case is made at the institution level, then the classification must first be reviewed by the warden and then forwarded for final review to the special monitoring staff at the correctional agency's headquarters. The initial classification of an inmate as a special monitoring case will be in effect pending the review. The review of a special monitoring classification decision should take place within sixty days and should determine whether or not a sound factual basis for the classification exists.
- If a special monitoring classification is overruled by the reviewing authorities (either the warden or the special monitoring staff at headquarters), then the institution's special monitoring coordinator should be informed immediately. The inmate's name should be removed from the special monitoring system and the special monitoring entry should be deleted from the automated inmate information system. Further, the inmate's central file should clearly reflect that the inmate has been removed from special monitoring status. Finally, the inmate must be notified in writing that he or she is no longer classified as a special monitoring case. The inmate should be requested to sign a notification form; if the inmate refuses to do so, a staff member witnessing this refusal should note the refusal on the form and sign it.

3. Review of Special Monitoring Status

- Except for inmates classified as special monitoring cases by virtue of being special witnesses or by virtue of serving federal sentences or out-of-state sentences, the appropriateness of continuing an inmate's special monitoring status should be considered at each program review. Specifically, case management staff should determine whether the special monitoring classification meets current criteria, whether there is sufficient documentation to support the assignment, and whether the reasons for the assignment are still valid.
- Staff should notify the institution's special monitoring coordinator in writing if they wish to recommend that an inmate's special monitoring classification be discontinued or modified. The recommendation should be routed from the special monitoring coordinator to the warden, and from the warden on to the special monitoring coordinator at the correctional agency's headquarters. Special monitoring status will remain in effect unless and until a recommendation for discontinuation or modification has been approved by the reviewing authorities.
- If an inmate is removed from special monitoring status, he or she should be notified, the central file amended, and the automated inmate information system corrected, in the same fashion as noted in "Assignments and Classification Procedures," above.
- Because participation in special witness programs is voluntary, an inmate classified as a special monitoring case by virtue of being a special witness may request removal from special monitoring at any time. All such inmate requests should be forwarded to the special monitoring coordinator at headquarters for action.
- Inmates serving federal or out-of-state sentences should be classified as special monitoring cases only during the duration of those sentences. If an inmate completes a federal or out-of-state sentence and then begins serving a separate in-state sentence, he or she should no longer be classified as a special monitoring case (unless other factors come to light that make it appropriate to place the inmate in a different special monitoring category once again).

4. Classification Appeals

- An inmate may appeal a special monitoring classification at any time. Ordinarily, the appeal of a special monitoring classification

should be made through the correctional agency's existing inmate grievance process.
- As an alternative to pursuing an appeal through the regular grievance process, inmates classified as special monitoring cases because they are special witnesses have the option of submitting their written appeals, as private correspondence, directly to the executive-level official at headquarters who has operational responsibility for the special monitoring program.
- Inmates in contract facilities typically do not have access to the regular inmate grievance process. They should submit their written appeals, as private correspondence, directly to the executive-level official at headquarters who has operational responsibility for the special monitoring program.

5. Classification of Recommitted Offenders

- Inmates who are in special monitoring status at the time of release from custody should retain that status in the event they are recommitted to the custody of the correctional agency. The special monitoring status should be reviewed, however, at the time of initial classification under the new commitment. If case management staff determines that reasons still exist for keeping the inmate in special monitoring status, they should follow the same review, authorization, and notification process outlined in "Assignments and Classification Procedures," above; if the staff determines that continuation of special management status is no longer warranted, they should follow the same review, authorization, and notification process outlined in "Review of Special Monitoring Status," above.

6. Special Monitoring Activities Clearance

- An inmate classified as a special monitoring case may not be transferred (except for medical or mental health emergencies), may not be given a temporary release, and may not participate in community activities, without prior clearance from the appropriate reviewing authority. Prior clearance from headquarters is required for any of the following actions involving an inmate classified as a special monitoring case:

—transfer to another facility within the system (other than a satellite camp)
—transfer to a contract facility or contract halfway house (for continued service of sentence)
—furlough
—escorted trips beyond commuting distance from an institution
—work release
—study release
—participation in community activities
—local escorted medical trips for inmates in the special witness category

The warden may give clearance for the following activities:
—transfer of inmates in the separation category from the institution's main facility to the same institution's satellite camp
—local escorted medical trips (other than special witness cases)
—local furlough medical trips (i.e., day trips) for out-of-state prisoners and separation cases

VII. SPECIAL HOUSING

Overview

There are three types of special housing units in which institutions may place inmates who may not be housed, for one reason or another, with the general inmate population: Disciplinary Detention Units, Administrative Segregation Units, and Special Management Units (also known as Control Units). The purposes, requirements, and operations of these types of units are quite distinct from each other.

Disciplinary detention units are used to confine inmates who have been determined by the disciplinary hearing officer to have committed serious violations of the correctional agency's policies and who warrant segregation from the general population for a specified period of time as a sanction. Administrative segregation units are used to house protective custody cases, inmates en route to other institutions (i.e., holdovers), inmates awaiting disciplinary hearings, and other inmates whose separation from the general population is necessary for the safety, security, or orderly operation of the institution. Unlike placement in a disciplinary detention unit, placement in an administrative segregation unit is nonpunitive in nature. Special Management Units provide ultra-secure housing and high-supervision programming exclusively for those inmates who, if confined in any less secure setting, would present the most extreme threats to others or to the orderly operation of the institution. Placement of an inmate in a Special Management Unit should be undertaken only in the rarest cases and must always be carefully documented. Special Management Units are necessary for the correctional agency to fulfill its obligation to provide safekeeping, care, and subsistence to those inmates who are extraordinarily violent or disruptive, while at the same time ensuring that those inmates will not be in a position to disrupt industrial, educational, and vocational training or other programs that serve the vast majority of inmates.

Each institution should have units for disciplinary detention and administrative segregation, or at least cells set aside for those purposes. Because of the heightened level of security, the longer term nature of confinement, and the rarity of use, however, Special Management Units are not appropriate for each institution. Instead, the correctional agency should designate one or more institutions as sites for Special Management Units.

1. Responsibility and Documentation

- Ordinarily, placement of an inmate in disciplinary detention is the responsibility of the institution's disciplinary hearing officer, with subsequent reviews of the inmate's status to be conducted by the institution's segregation review official. Placement of an inmate in administrative segregation is the responsibility of the warden or the warden's designee (normally the shift commander), with subsequent reviews of the inmate's status to be conducted by the institution's segregation review official. Placement of an inmate in special management is made by the appropriate executive-level official at the correctional agency's headquarters (a deputy commissioner or deputy director with responsibility for security and correctional management), on the recommendation of the warden and with the advice of the headquarters' hearing administrator. Subsequent reviews of the inmate's status would be made by the hearing administrator and ratified by the deputy commissioner or deputy director.
- Each inmate in special housing status must be visited by a senior correctional supervisor and a qualified health care official every day to monitor each inmate's adjustment and to ensure that

policies with respect to inmates in special housing are being carried out properly.
- Copies of all documentation relating to placement of an inmate in special housing status—notices of hearings, segregation orders, progress reports, orders for digital or simple instrument searches to be carried out on the inmate, etc.—must be placed in the inmate's central file. Logs must be maintained in each special housing unit, in which all significant events should be recorded (including confinement of new inmates, release of inmates, temporary movement of inmates to the hospital or other sites in the institution, hearings, disturbances and misconduct, visitors, etc.).

2. Placement of Inmates in Disciplinary Detention

- Ordinarily, an inmate may be placed in disciplinary detention only by order of the disciplinary hearing officer, following a hearing at which the inmate was found to have committed an infraction in the greatest, high, or moderate category or a repeated offense in the low-moderate category (see Part VIII). The disciplinary hearing office may order placement in disciplinary detention only when other available sanctions are not deemed to be sufficient punishment or to be an effective deterrent for regulating an inmate's behavior.
- Under the following extreme situations, inmates may be placed in disciplinary detention before the disciplinary hearing officer can schedule a hearing and render a decision: (a) if an inmate in administrative segregation is causing a serious disruption and needs to be moved to a more secure setting, or (b) if the Medical Department advises that an inmate who would ordinarily be housed in the institution hospital for mental or physical treatment requires a higher level of security than the hospital is able to provide.
- The segregation review official must conduct a hearing on each inmate who spends seven consecutive days in disciplinary detention, formally review the inmate's case, and document the inmate's status. Thereafter, the segregation review official should review the inmate's case (in the inmate's absence) every seven days and conduct a formal hearing on the inmate's case at least once every thirty days. A mental health assessment of the inmate should be conducted as part of each thirty-day hearing (see "Health Care,"

below). The inmate should appear before the segregation review official at the hearings, but may waive the right to do so by signing a formal waiver. If the inmate refuses to attend the hearing and also refuses to sign the waiver, the staff member who advised the inmate of the hearing should indicate the inmate's refusals in a signed memorandum and a second staff member should countersign the memorandum as a witness.
- The segregation review official, in consultation with the chief of security and the disciplinary hearing officer, may release an inmate from disciplinary detention before the initially imposed sanction expires if he or she determines that detention is no longer necessary to regulate the inmate's behavior or fulfill the purposes of punishment and deterrence. The segregation review officer, however, may not increase the duration of any previously imposed sanction.

3. Placement of Inmates in Administrative Segregation

- The warden is authorized to place inmates in administrative segregation, although ordinarily the warden should delegate that authority to the shift commander. Inmates may be placed in administrative segregation when they: (a) are in holdover status (i.e., an inmate in transit to another institution); (b) are new commitments, pending classification; (c) are likely to pose a serious threat to themselves, staff, other inmates, property, or the security and orderly operation of the institution, if they were among the general population; (d) are awaiting a hearing on charges of violating correctional agency regulations; (e) are being investigated for a possible violation of correctional agency regulations; (f) are being investigated for possible criminal violations; (g) are awaiting transfers; (h) have requested placement in administrative segregation for their own protection; (i) have been determined by staff to require placement in administrative segregation for their own protection; and (j) have completed confinement in disciplinary detention but should not be returned to the general population.
- Protection cases may include: (a) victims of inmate assaults; (b) inmate informants; (c) inmates who have been pressured by other inmates to participate in sexual activity; (d) inmates who request protection and claim to be former law enforcement officers, informants, or others in sensitive law enforcement positions,

regardless of whether official information is available to verify the claim; (e) inmates who refuse to enter the general population because of alleged pressures from other unidentified inmates; (f) inmates who refuse to enter the general population, even if they refuse to provide and staff cannot determine a reason for this refusal; and (g) inmates about whom staff have good reason to believe are in serious danger of bodily harm.
- Staff should prepare an administrative segregation order outlining the reasons for an inmate's placement in administrative segregation. A copy of this order should be given to the inmate within twenty-four hours or his or her placement in administrative segregation, unless doing so would compromise institution security. Copies of the order should also be given to the inmate's unit team and placed in the inmate's central file. Administrative segregation orders are not required when the inmate is placed in administrative segregation solely because he or she is on holdover status.
- The segregation review officer should conduct a record review within three working days of an inmate's placement in administrative segregation to determine if this placement remains appropriate. He or she should conduct a formal hearing to review the status of each inmate who spends seven continuous days in administrative segregation. Thereafter, the segregation review official should conduct a record review (in the inmate's absence) on a weekly basis and should hold a formal hearing at least once every thirty days. A mental health assessment should be conducted as part of each thirty-day hearing (see "Health Care," below). The inmate should appear at formal hearings before the segregation review official but may waive the right to do so by signing a written waiver. If the inmate refuses to sign the waiver, this should be noted in a memorandum signed by the responsible staff member and countersigned by a second staff member who witnessed both the inmate's refusal to attend the hearing and the refusal to sign the waiver. The inmate should receive a written copy of the segregation review official's findings in each review or hearing, unless releasing the findings to the inmate might compromise institution security.
- Whenever it is necessary for an inmate to remain in administrative segregation for more than ninety days, the matter should be referred to the appropriate executive-level official at the correctional agency's headquarters for review and disposition. The institution should provide all documentation to substantiate

recommendations that the inmate not be returned to the general population.
- Administrative segregation should be used only for short periods of time, except in cases where it is documented that an inmate requires long-term protection or where there are exceptional circumstances (such as security concerns or the need to conduct complex investigations). The segregation review official must release an inmate from administrative segregation once the reasons for the placement cease to exist.

4. Placement of Inmates in Special Management Units

- The warden may initiate the process for placing an inmate in a Special Management Unit by submitting a written recommendation to the executive-level official at the correctional agency's headquarters (the deputy commissioner or deputy director in charge of security and correctional management issues). The referral should include a memorandum stating the basis for the warden's recommendation, copies of all disciplinary reports and investigative materials related to the specific acts that prompted the recommendation, a copy of the presentence investigation report, and copies of the most recent and fully updated progress reports (including information on misconduct incidents), medical reports, and mental health reports.
- In recommending an inmate for special management status, the warden may consider the following factors: (a) any incident during confinement in which the inmate has caused injury to others; (b) any incident in which the inmate has made threats against the life or well-being of others; (c) an incident in which the inmate has been involved in disrupting the orderly operation of a correctional facility; (d) an escape or attempted escape from a correctional facility (especially in cases where the escape or escape attempt has involved hostage taking or the use of deadly weapons; less threatening escape attempts may only warrant redesignation of the inmate to a more secure facility); and (e) the nature of the offense for which the inmate was committed (but only when considered in combination with one or more of the other factors; in and of itself, the offense for which an inmate was committed would not warrant placement of the inmate in a Special Management Unit).

- Inmates who show evidence of significant mental disorder or major physical disabilities, as documented in a mental health evaluation or physical evaluation, may not be recommended for placement in a Special Management Unit.
- If the deputy commissioner or deputy director at headquarters concurs with the warden's recommendation for placing an inmate in the Special Management Unit, he or she should direct the headquarters official designated as the hearing administrator to review the referral materials and conduct a hearing on the appropriateness of the recommendation.
- The hearing administrator should contact the warden to arrange a time and date for the hearing, which ordinarily should be held at the institution. The hearing administrator should also prepare a Notice of Special Management Unit Hearing that would explicitly indicate all disciplinary actions, evidence, and other matters to be considered at the hearing. The notice should be given to the inmate at least twenty-four hours prior to the hearing, along with a copy of the correctional agency's current policy on Special Management Units. The date and time that the inmate receives the notice should be recorded on the institution's copy of the notice. If the inmate is illiterate, the staff should read the notice to the inmate and should explain its significance.
- The hearing administrator should provide the inmate with the service of a full-time staff member as his or her representative at the hearing, if the inmate so desires, who may assist the inmate, contact witnesses, and present favorable evidence at the hearing. A staff representative should be afforded reasonable time by the hearing administrator to meet with the inmate and interview witnesses.
- The inmate may select a staff representative from the local institution. If the selected staff member declines or is unavailable, the inmate has the option of selecting another representative, waiting a reasonable period of time (as determined by the hearing administrator) for the staff member to return from an absence, or proceeding without a staff representative. The hearing administrator should document in the record of the hearing an inmate's request for, or refusal of, staff representation.
- The inmate has the right to be present throughout the hearing, except where institutional security or good order may be jeopardized. The hearing may be conducted in the inmate's absence, if the inmate refuses to appear. An inmate's refusal to appear, or

other reason for nonappearance, must be documented in the record of the hearing. An inmate who does not appear at a hearing retains the right to have a staff representative and witnesses appear on his or her behalf.
- The inmate (or the inmate's staff representative) is entitled to present documentary evidence and have witnesses appear at the hearing, provided that calling witnesses does not jeopardize institutional security or safety, and also provided that the witnesses are available at the institution where the hearing is being conducted. In cases where witnesses cannot appear, written statements from witnesses may be submitted.
- Evidence presented at hearings should concern the issue of whether an inmate could function in a general prison population without posing a threat to others or to the orderly operation of the institution. The hearing administrator has no authority to reverse prior findings of disciplinary violations, and therefore no evidence should be submitted for the purpose of challenging those findings.
- Following the hearing and a thorough review of all material related to the recommendation for placement of an inmate in special management status, the hearing administrator should prepare a written decision endorsing or rejecting the recommendation. The written decision should summarize the hearing and all information and evidence on which the decision was based, and it should indicate the specific reasons for the decision.
- The hearing administrator's decision, whether for or against placement of the inmate in a Special Management Unit, should be forwarded (along with supporting documentation) to the deputy commissioner or deputy director within twenty working days after the conclusion of the hearing. The inmate should receive a copy of the decision, unless the release of certain portions of it could pose a threat to safety or security, in which case that limited information may be withheld. The inmate should be advised that the hearing administrator's decision will be referred to the deputy commissioner or deputy director for final disposition, and that the inmate has five working days to file an appeal of the decision with the deputy commissioner or deputy director. The date and time of the inmate's receipt of the hearing administrator's decision should be recorded on the institution's copy of the decision.
- The final decision on whether or not an inmate should be placed in special management may not be delegated by the deputy commissioner or deputy director. That official should review the

hearing administrator's decision, the supporting documentation, and the appeal (if any) submitted by the inmate and then render a final decision within thirty working days after the hearing administrator's decision was received. The deputy commissioner or deputy director's decision should be forwarded to the warden at the institution to which the inmate will be transferred (if applicable), the warden who recommended placement in special management, the hearing administrator, and the inmate.
- If the deputy commissioner or deputy director approves the inmate for placement in a Special Management Unit, the inmate may appeal the decision (via the grievance process) directly to the chief executive officer of the correctional agency (i.e., the commissioner or director). The inmate should be advised of appeal rights at the same time he or she is informed of the decision.
- Pending transfer to a Special Management Unit, the inmate should be considered to be in holdover status, and confined in administrative segregation. If the transfer is delayed, follow-up examinations may be required prior to the eventual transfer.
- The warden at the transferring facility should ensure that the results of the inmate's most recent physical examination (to be conducted not more than thirty days prior to transfer) and most recent mental health examination (to be conducted no more than ninety days prior to transfer) accompany him or her to the Special Management Unit.
- Unit staff should monitor the inmate's adjustment to the Special Management Unit on a daily basis.
- The Special Management Unit Team (consisting of the unit manager, the case manager, the counselor, education staff member assigned to the unit, and other members designated by the warden) should meet with the inmate once every thirty days to assess the inmate's progress. The unit team should review the inmate's daily activities, a psychological assessment report, and a personal interview with the inmate, and make a recommendation as to the inmate's readiness for release from the Special Management Unit. The inmate is required to attend the meeting in order for the previous month's confinement in the Special Management Unit to be credited toward the estimated duration of his confinement there.
- The unit team's recommendation should be submitted to the warden for review.
- The Deputy Director or Deputy Commissioner should review the inmate's status in special management at least once every sixty to

ninety days, to determine the inmate's readiness for release from the unit. In addition to recommendations from the unit team and the warden, the Deputy Commissioner or Deputy Director should consider the following factors: (a) the inmate's relationship with other inmates and staff members, demonstrating whether the inmate is able to function in a less restrictive environment without posing a threat to others or to the orderly operation of the institution; (b) the inmate's involvement in work assignments and recreational activities; (c) the inmate's adherence to institutional guidelines and the correctional agency's regulations and policies; (d) the inmate's personal grooming and the cleanliness of his or her quarters.

- A copy of the written decision of the Deputy Commissioner or Deputy Director should be forwarded to the inmate. The inmate should be advised that he or she has thirty calendar days from the receipt of an adverse decision to submit an appeal, via the regular inmate grievance procedure, directly to the Commissioner or Director of the correctional agency.
- Only the Deputy Commissioner or Deputy Director with responsibility over special management activities, or the correctional agency's chief executive officer (i.e., the Commissioner or the Director) may release an inmate from the Special Management Unit. If the inmate is released from special management, he or she should be transferred to an appropriate facility for completion of sentence.

5. Programming and Conditions of Confinement

- In general, conditions of confinement in the three types of special housing will be fairly similar, except for the fact that fewer amenities and services may be available to inmates in disciplinary segregation than in the other types because it is intended to be punitive and short-term.
- The warden must ensure that living conditions in all special housing areas meet basic levels of decency and humane treatment. Living conditions may not be modified for the purpose of compelling acceptable inmate behavior. Quarters must be well-ventilated, adequately lighted, appropriately heated, and maintained in a sanitary condition at all times. All cells must be equipped with beds, which may be securely fastened to the floor or to the cell wall. The number of inmates confined to a cell must

not exceed the number for which the space was designed, unless the warden determines there is a pressing need for exceeding occupancy limits.
- Inmates in special housing may not be segregated without clothing, mattress, blankets, and pillow, except when prescribed by the medical officer for medical or mental health reasons (for example, in cases where an inmate is so seriously disturbed that he or she is likely to destroy clothing or bedding or create a disturbance that would be detrimental to others). Inmates in special housing may wear normal institution clothing, but may not have a belt. Cloth or paper slippers may be substituted for shoes, at the discretion of the warden. To the greatest extent practicable, inmates in special housing should be provided with the same opportunity for the issue and exchange of clothing, bedding, and linen, and for the laundering of clothing, as inmates in the general population. Exceptions to this procedure may be permitted only at the warden's direction, in which case the reasons for the exception must be carefully documented in the unit log.
- Inmates should be provided with toilet tissue, a wash basin, a toothbrush, eyeglasses, shaving utensils, etc., as needed. Unless it would present an undue security hazard, as determined by the warden, inmates in special housing should be permitted to shower and shave at least three times a week (razors should be controlled by Special Housing Unit staff, and only disposable razors should be used). Where practicable, barbering and hair care services should be provided.
- Inmates in special housing should be given nutritionally adequate meals, ordinarily from the menu of the day for the institution. When safety or security reasons dictate, or when throwing or other misuse of food has been documented, inmates in disciplinary detention may be given nutritionally balanced food loaves instead of meals from the institution's daily menu.
- Inmates in disciplinary detention and administrative segregation should be permitted a minimum of five hours recreation and exercise out of their cells every week. Recreation may be suspended at the warden's discretion, however, if there are compelling safety or security reasons for doing so, or if an inmate has been denied recreation as a punitive measure (for a period not to exceed one week) following a disciplinary hearing and a recommendation by the Disciplinary Hearing Officer. If weather and resources permit, the opportunity for outdoor recreation should be provided. Normally, each recreation period should be one-hour

long, and should be given on five different days; if necessary, however, inmates may be given two half-hour recreation periods in a day, provided the total recreation time for the week equals five hours.

Inmates in special management status should receive a minimum of seven hours of out-of-cell recreation and exercise per week, and should be provided with games and exercise equipment (not to include weight training equipment). Outdoor recreation should be permitted if practicable, and foul weather gear may be provided to inmates wishing to take outdoor recreation during inclement weather. Special management inmates ordinarily would take recreation individually. They may not take recreation in groups, unless the warden recommends that group recreation be permitted, and the warden's recommendation is approved by the executive-level official at headquarters (i.e., the Deputy Commissioner or Deputy Director) responsible for custody and operations.

- Personal property of inmates in disciplinary detention should be impounded for the duration of their confinement in that status, and they are not eligible for commissary privileges. Inmates in disciplinary detention, however, may be permitted to possess a reasonable amount of nonlegal reading material (generally from the institution's circulating library), legal materials in accordance with agency policy, and religious scriptures of their particular faiths.
- Unless compelling safety or security reasons make it impracticable, inmates in administrative segregation should be permitted to possess a reasonable amount of personal property, which may include religious literature and authorized devotional objects (such as medals and headgear), legal materials, magazines and newspapers, mail, personal hygiene items (except razors), photo albums, shower slippers, tennis shoes (except for those containing steel shanks), snack foods, fruit, powdered soft drinks, coffee, cigarettes (unless the unit is a smoke-free area), drinking cups, sugar cubes, chewing tobacco, chewing gum, stationery, postage stamps, wedding band, watch, radio with earplug, and eyeglasses. Inmates in administrative segregation should also have commissary privileges.
- Inmates in Special Management Units should be permitted to possess personal property, provided it does not exceed allotted space in the cell, and is limited to specific items approved for special management inmates. Typically, items made of glass or metal would not be allowable as property for special management

inmates. In addition, special management inmates may retain up to three cubic feet of legal materials in their cells; excess legal materials, at the inmate's option, may be destroyed, placed in storage at another location, or mailed home. Special management inmates may also have commissary privileges, although the types of commissary items available to them may be limited by the warden in accordance with safety or security needs.
- Although inmates in disciplinary detention are not eligible for most program privileges, other than religious guidance and medically necessary counseling, inmates in administrative segregation should have the opportunity to participate in program activities available to the general population. If consistent with available resources and the security needs of the unit, these program opportunities for administrative segregation inmates should include (but not be limited to) participation in an education program, library services, social services, counseling, and religious guidance.
- Comprehensive and closely supervised programming should be available to special management inmates. Such programming may be suspended, with the warden's written authorization, only for compelling reasons of security or safety, and only if the reasons have been carefully documented. The Special Management Unit team should include a case manager, responsible for coordinating all regular case management functions, and should also include a counselor who can handle inmates' concerns and requests, provide grievance forms, and be available for consultation and counseling.

The warden should assign a member of the Education Department to the Special Management Unit on at least a part-time basis to assist in developing educational programs to fulfill each inmate's academic needs. Study courses should be provided at all levels, including Adult Basic Education, General Equivalency Diploma programs, correspondence courses, and selected areas of special interest.

Staff may assign inmates to a work assignment, such as range orderly. If an industrial program is established in a Special Management Unit, it should be supervised by an industry foreman and participating inmates should earn industrial good-time credits and industrial pay (consistent with applicable regulations).

Special management inmates may possess religious materials, the type and amount of which may be limited only by security considerations or housekeeping regulations (food, for example, may not be stored in an inmate's cell for use as sacraments because of the sanitation problems that would result). Institution or contract chaplains should visit the Special Management Unit at least once a week and may make additional visits as needed. While special management inmates may pray or worship individually or with a chaplain, religious assemblies or group meetings would pose security problems in the Special Management Unit and should not be permitted.

6. Health Care

- Inmates in special housing units must be permitted to continue taking prescribed medication and must be visited by a member of the Medical Department every day (including weekends and holidays). Special housing inmates must undergo a mental health assessment (including a personal interview) after thirty days and at subsequent intervals of thirty days (although such extended confinements, while typical for special management cases, would be unusual for disciplinary detention and administrative segregation cases). Mental health assessments should be submitted in writing to the segregation review official and should address issues such as inmates' adjustment to surroundings and any threats the inmates might pose to themselves, to the staff, or to other inmates.

7. Legal Matters and Entitlements

- Inmates in disciplinary detention and administrative segregation should have access to legal materials from the inmate law library. Legal visits, legal telephone calls, and legal correspondence must be permitted in accordance with the correctional agency's policies. Social visiting may be restricted if substantial reasons exist for doing so, and staff should make reasonable efforts to notify approved social visitors of those restrictions in order to spare them disappointment and unnecessary inconvenience. Telephone privileges for inmates in disciplinary detention may be limited, except for calls to their attorneys of record. Inmates in administrative segregation, however, should have telephone privileges in accordance with the correctional agency's established policy. Inmates in disciplinary detention and administrative segregation should have regular correspondence privileges, in accordance with the correctional agency's established policy.

- A basic inmate law library should be established for each Special Management Unit. Inmates in special management should have access to this library, upon request and in rotation. Consistent with security considerations, this library should include basic legal reference books, a table and chair, a typewriter, and stationery. Reference books available in the main law library, but not in the Special Management Unit's law library, may be obtained for an inmate upon request. If legal reference books are abused by inmates, inmates may be required to use them under closer supervision (e.g., in their cells). Abuse of materials in the Special Management Unit's law library (books, typewriters, etc.) may result in limitations being placed on the use of legal materials, provided that decision is made by the warden and documented in writing.
- Social visits for inmates in the Special Management Unit should be conducted in a controlled, noncontact visiting area, separated from the institution's regular visiting facilities. Four hours per month visiting time are generally allotted to special management inmates. All visitors must be on the inmate's approved visiting list. Limits may be placed on the number of visitors an inmate may receive, on the number of visits in excess of four per month, the number of visiting hours in excess of the four per month normally allotted, and the number of consecutive hours of visiting on a given day, in order to be consistent with available resources and institutional security and good order. Legal visits must be permitted in accordance with the correctional agency's established policy.
- Special management inmates may have telephone and correspondence privileges in accordance with the correctional agency's established policies.

8. Additional Special Management Procedures

- Upon admission to the Special Management Unit, inmates should be advised of the estimated duration of their confinement. This may range from one month to an indefinite number of months and is based largely on the nature of the act or acts that resulted in the inmate's placement in special management status and on the inmate's behavior while in administrative segregation awaiting transfer to the Special Management Unit. Incoming special management inmates should also be advised that their confinement in the Special Management Unit may be shortened or extended, depending on their behavior in the unit.

- Staff should provide incoming special management inmates with a summary of the unit's guidelines and disciplinary procedures and a list of authorized personal property items, explain activities in the Special Management Unit and the expectations for the inmate's involvement in those activities, and advise inmates of the criteria for release from the unit.
- The warden may order a digital or simple instrument search of all new admissions to the Special Management Unit, and of any inmate received at or returned to the Special Management Unit following contact with the public. The decision for ordering a digital or simple instrument search may not be delegated below the level of acting warden, and the order must be in writing, signed by the warden or acting warden. The option to conduct this procedure is required because some inmates transport serious contraband, such as hacksaw blades, in their body cavities; undetected, such contraband poses a serious threat to institution security, good order, and the personal safety of staff and inmates. The threat of this type of contraband is heightened in settings such as Special Management Units.
- Unless the suspected contraband would not be detectable by an x-ray, an inmate may submit a written request that an x-ray be taken in lieu of the digital or simple instrument search. The warden will rule on the request, after consultation with the institution's chief medical officer or acting chief medical officer.
- For further information on proper procedures for digital, simple instrument, or x-ray searches, see Part II.

VIII. INMATE DISCIPLINE

Overview

To ensure a safe and orderly environment for staff and inmates, it is necessary for institution staff to impose discipline on those inmates who violate regulations of the correctional agency. Administration of discipline in a prison setting must be governed by the following guiding principles:

- Disciplinary policy should apply to all inmates in the custody of the correctional agency, including committed inmates; pretrial detainees; inmates out on writs, escorted trips, furloughs, and out-of-state or federal inmates serving on a contract basis in one of the

correctional agency's facilities. Inmates in nonstate facilities (such as contract halfway houses) are not subject to the correctional agency's disciplinary policy *per se*, but community program managers are authorized to take disciplinary actions as specified in regulations governing community programs.
- Only staff members (i.e., employees of the correctional agency) may take disciplinary action (unless a disciplinary matter is referred to other appropriate law enforcement agencies, such as the state police or prosecutor's office). Permitting discipline to be meted out by inmate "trustys," "kangaroo courts," "tenders," "gang bosses," or anyone else is unprofessional, dangerous, potentially disruptive of institutional security and good order, and a gross violation of sound correctional practice.
- Staff should take disciplinary action at such times and only to the degree necessary to manage an inmate's behavior within the correctional agency's regulations and institution guidelines and to promote a safe and orderly institution environment.
- Staff should impose discipline in a completely impartial and consistent manner. Disciplinary action must never be capricious or retaliatory and must always adhere strictly to institution regulations governing the investigation of infractions and the administration of discipline.
- Staff must not impose or permit corporal punishment of any kind.
- If it appears at any stage in the disciplinary process that an inmate is mentally ill, the staff must refer the inmate to a mental health professional for determination of whether the inmate is responsible for his or her conduct or is incompetent. Disciplinary action may not be taken against inmates whom mental health staff have determined to be incompetent or not responsible for their actions.
- Any staff member who serves as a disciplinary hearing officer or a member of a Unit Discipline Committee must successfully complete the training program in the correctional agency's disciplinary practices. Agency counsel must certify the competence of each disciplinary hearing officer to implement agency policies on inmate discipline.

1. Levels of Severity

- To ensure that the level of discipline is commensurate with the severity of the offense, the correctional agency should establish a

list or schedule of offenses, ranked by severity or seriousness, that would be correlated to a list or schedule of appropriate sanctions. The severity of an offense would also correlate to the level of disciplinary review it would warrant. Less serious infractions could be disposed of informally by the shift commander; more serious infractions would have to be referred to the Unit Discipline Committee and, if appropriate, the disciplinary hearing officer. The most serious infractions, if they involve possible criminal wrongdoing, would be referred to appropriate outside law enforcement agencies.
- Four main categories of offenses may be identified as rising to the level of warranting disciplinary attention: prohibited acts of the *greatest severity, high severity, moderate severity,* and *low severity.*

Greatest Severity

- Acts that typically would be categorized as being of the greatest severity would include killing; physical assault with intent to injure; sexual assault; armed attack on the institution's secure perimeter; escape from a secure institution; escape from a minimum-security institution involving violence; escape from escort; the possession, manufacture, or introduction of a weapon (including guns or other firearms, sharpened instruments, knives, dangerous chemicals, explosives, or ammunition); rioting; encouraging others to riot; hostage taking; the possession, manufacture, or introduction of hazardous tools (such as those that could be used in an escape attempt or as weapons); the possession, introduction, or use of narcotics (except as prescribed by medical staff); refusal to provide a urine sample or participate in other types of drug testing; interference with a staff member in the performance of duties (if the interference occurs in connection with another act of greatest severity); and conduct that disrupts or interferes with the security or orderly management of an institution (if the conduct occurs in connection with another act of greatest severity).

Sanctions appropriate for infractions in the greatest severity category would include one or more of the following: recommendation of rescission or postponement of parole date; forfeiture of up to 100 percent of statutory good time and termination or disallowance of extra good time; recommendation of disciplinary transfer; disciplinary detention of up to sixty days; monetary restitution; and loss of privileges (as a supplemental punishment only—this cannot be the only sanction meted out at this level).

High Severity

- Offenses in the high severity category would include escape or attempted escape from unescorted community programs and activities and from minimum-security facilities (unless violence was committed in doing so); fighting; threatening another person with bodily harm; extortion, blackmail, or demanding or receiving payment of anything of value in return for protection; engaging in sexual acts; making sexual proposals or threats; wearing of a mask or other form of disguise; possessing any unauthorized locking devices, keys, or lock picks; tampering with, blocking, damaging, or otherwise interfering with keys or other locking devices, mechanisms, or procedures; adulterating any food, drink, or property; possessing officers' clothing or other types of staff clothing; engaging in or encouraging a group demonstration; participating in a work stoppage or encouraging others to do so; introducing alcohol into the institution; giving or offering a bribe to an official or staff member; giving or receiving money for the purpose of introducing contraband, or for any other prohibited activity; destroying, altering, or damaging government property or the property of another person (if that property is worth more than $100); destroying, altering, or damaging life-safety devices, such as fire alarms (regardless of financial value); stealing; engaging in the martial arts, boxing, wrestling, or other forms of physical encounters or military exercises; being in an unauthorized area with a member of the opposite sex, without staff permission; making, possessing, or using intoxicants; refusing to participate in breathalyzer test or other tests intended to detect alcohol use; physical assaults that are less serious in nature from assaults categorized as being offenses of greatest severity; interfering with a staff member in the performance of duties (if the interference occurs in connection with another act of high severity); and engaging in conduct that disrupts or interferes with the security or orderly management of the institution (if the conduct occurs in connection with another act of high severity).

Sanctions for infractions in the high severity category would include recommendation of rescission or postponement of parole date; forfeit of up to 50 percent or up to sixty days of statutory good time (whichever is less), and/or termination or disallowance of extra good time; recommendation of disciplinary transfer; disciplinary detention of up to thirty days; monetary restitution; withholding of statutory good time; loss of privileges (such as commissary, movies, recreation); transfer to different quarters within the institution; removal from program or group activity; loss of job; impoundment of personal property; and restriction to quarters.

Moderate Severity

- Offenses in the moderate severity category would include indecent exposure, misusing authorized medication, possessing money or currency (unless specifically authorized, or in excess of the authorized amount); loaning property or anything of value to another inmate for profit or increased return; possessing any items that are not authorized or were not issued through authorized channels; refusing to work; refusing to accept a program assignment; violating a condition of a furlough or community program; having an unexcused absence from work or any assignment; failing to perform work as instructed by a supervisor; being insolent toward a staff member; lying or giving false statements to a staff member; participating in an unauthorized meeting or gathering; being in an unauthorized area; failing to follow safety or sanitation regulations; using any equipment or machinery without authorization or contrary to instructions or posted safety standards; failing to stand count; interfering with the taking of count; gambling; preparing or conducting a gambling pool; possessing gambling paraphernalia; having unauthorized contacts with the public; giving money or anything of value to another inmate or anyone else without authorization from staff; receiving money or anything of value from another inmate or anyone else without authorization from staff; and failing to keep one's person and one's quarters in accordance with posted standards for sanitation and tidiness.

Sanctions appropriate for violations in the moderate severity category would include recommendation for rescission or postponement of parole date; forfeiture of up to 25 percent or up to thirty days of earned statutory good time (whichever is less), and/or termination or disallowance of extra good time; withholding of statutory good time; recommendation of disciplinary transfer; disciplinary detention of up to fifteen days; monetary restitution; loss of privileges (commissary, movies, recreation, etc.); transfer to different quarters within the institution; removal from program activities or group activities; loss of job; impoundment of personal property; restriction to quarters; and extra duty.

In addition, there are several offenses that normally would be considered moderate severity but could also be considered of greater severity if carried out in connection with other acts of greater severity. Refusal to obey any order from any staff member; possessing, manufacturing, or introducing a nonhazardous tool or other nonhazardous contraband; and counterfeiting, forging, or making unauthorized reproductions of any document, article of identification, money, security, or official paper normally would fall in the category of moderate severity. If refusing to obey an order, however, could contribute to a riot, then it would be considered an offense of greatest severity; if failing to obey an order contributes to a fight, then it would be considered an offense of high severity. Similarly, if an inmate manufactures a nonhazardous tool or forges a document for use in an attempt to escape from a secure facility, those offenses should be considered of greatest severity.

Low Severity

- Offenses in the low severity category would include possessing property belonging to someone else; possessing an unauthorized quantity of otherwise authorized clothing; malingering or feigning illness; smoking in a prohibited area; using abusive or obscene language; tattooing or self-mutilation; conduct with a visitor that violates regulations of the correctional agency; conducting a business; and engaging in unauthorized physical contact (such as kissing or embracing).

Sanctions appropriate for violations in the low severity category would include withholding statutory good time; loss of privileges; monetary restitution; transfer to different quarters in the institution; removal from program activity or group activity; loss of job; impoundment of personal property; restriction to quarters; extra duty; reprimand; and a warning.

In addition, there are several offenses that normally would be considered low severity, but could also be considered of greater severity if carried out in connection with other acts of greater severity. For example, the unauthorized use of mail or telephone privileges normally would be considered a low severity offense, but this would be considered of greatest severity if the misuse of mail or telephone privileges were undertaken to help plan an armed attack on the secure perimeter.

Meting Out Sanctions

- Within broad guidelines, disciplinary hearing officers should exercise discretion when meting out sanctions, taking into consideration all relevant factors (including whether the inmate has committed similar acts in the past) and assessing the severity of the offense. While authorized to disallow up to 100 percent of good conduct time credit available for a given year as a sanction for greatest severity offenses, the disciplinary hearing officer ordinarily would disallow between 50 percent and 75 percent—depending on the seriousness of the specific offense. Although both serious assaults and possession of marijuana would be offenses of greatest severity, the disciplinary hearing officer would likely disallow more good time for the former offense than for the latter. Similar discretion should be used when imposing sanctions in other categories.

2. Notifying Inmates of Disciplinary Practices

- All new inmates should be given written notification when they enter the institution of rules and regulations they are responsible for observing, of possible sanctions for violating rules and regulations, and of the disciplinary process that would be followed if they were accused of violating rules and regulations. If a significant part of the inmate population speaks a particular foreign language, printed materials should be made available in that language. Otherwise, translators should assist non–English speaking inmates in understanding their obligations and the disciplinary process.

3. Incident Reports and Staff Investigations

- Staff who note or suspect infractions have occurred should document the infractions in an incident report and submit the incident report to the shift commander. The incident report should specify the prohibited act allegedly committed by the inmate,

include all details about the incident known to the writer of the report (except for confidential information, which should not be included in an incident report), and indicate if any immediate action was taken in response to the incident (such as confiscating contraband or using force to stop an escape attempt). Each incident report must be signed and dated by the staff member who writes it.
- The shift commander has the authority to dispose informally of infractions in the low severity or moderate severity categories. A record of informal resolutions, giving the inmate's name, the inmate's registration number, the subject of the informal resolution, and the agreed-upon disposition, should be maintained by the chief of security. The disciplinary process may be suspended for up to two weeks following the submission of an incident report to a shift commander to work out and implement an informal resolution. If, at the end of the two-week period, an informal resolution has not been worked out or implemented successfully, then the disciplinary process would resume at the point it was discontinued.
- If an informal resolution of a low severity or moderate severity offense is inappropriate or unsuccessful, the shift commander should forward the incident report to the Unit Discipline Committee.
- The shift commander is not authorized to dispose informally of offenses in the high severity or greatest severity categories, and is required to refer such matters to the Unit Discipline Committee.
- Investigations should be conducted into incidents that cannot be disposed of informally. Normally, staff should commence an investigation within twenty-four hours of a reported violation, and the investigation should be completed within twenty-four hours. The investigating officer should be a supervisory-level staff member (ordinarily at the shift commander level or above). Neither the staff member who reported the incident being investigated nor any other staff member who may have been involved in the incident may be assigned as the investigating officer.
- The investigating officer normally should begin by interviewing the inmate accused of committing the prohibited act. The investigating officer should give a copy of the incident report to the inmate and document the date and time when the incident report was delivered to the inmate. If the inmate does not receive a copy of the incident report at the beginning of the investigation, then the reason for this must be stated in the investigative report.

The investigating officer should also advise the inmate of his or her right to remain silent at all stages of the disciplinary process and document that this information had been given to the inmate. The inmate should also be informed that, while his or her silence may be cause for adverse inferences at any stage in the institutional disciplinary process, silence alone may not be used to support an adverse finding.

- To the best extent practicable, statements by the inmate offering rationales for his or her conduct or for the charges should be investigated. For example, if an inmate who has received an incident report based on a positive urine test claims that the test result was affected by prescribed medication that he or she was taking, then the investigating officer should contact appropriate staff in the Medical Department for possible corroboration of the inmate's claim.
- The investigating officer should interview everyone with direct and relevant information about the incident and summarize their statements in the investigative report. In particular, the investigative officer should interview the staff member who submitted the incident report to obtain a firsthand account and provide additional clarification. Also, the investigative officer should record the disposition of evidence in the investigative report. Finally, the investigative report may include the investigating officer's comments on the inmate's prior record and behavior while in custody, analysis of any conflict or contradiction between witnesses, and conclusions as to what actually happened in the incident.
- When completed, the investigative report should be submitted, as appropriate, to either the Unit Discipline Committee or the disciplinary hearing officer.
- The inmate should not receive a copy of the investigative report. If the case is forwarded to the disciplinary hearing officer, however, the disciplinary hearing officer should make a copy of the report and other relevant materials available to the inmate's staff representative, for use in any presentation on the inmate's behalf.
- If it becomes apparent during an investigation that an incident may become the subject of a criminal prosecution, the investigation should be suspended immediately. To ensure that the staff does not jeopardize a criminal investigation, staff may not question the inmate until police detectives, prosecuting attorneys, or other appropriate law enforcement officials have completed their own interviews or until the agency responsible for the criminal

investigation advises the institution that it may proceed with its own questioning of the inmate. The institution, however, is not required to await the outcome of a criminal prosecution before taking disciplinary action against an inmate, unless the investigating agency or prosecutor's office has requested such a delay. An inmate who is the subject of a criminal prosecution for misconduct within the prison should not be transferred to a facility outside the jurisdiction of the trial court without the consent of the prosecutor's office.

4. Unit Discipline Committee

- Cases that cannot be disposed of by the shift commander must be referred to the Unit Discipline Committee. The Unit Discipline Committee holds initial hearings concerning charges of inmate misconduct following the completion of the investigative report and may impose minor sanctions (i.e., sanctions *other than* recommending rescission or postponement of parole date, forfeiture of earned statutory good time, termination or disallowance of extra good time, disciplinary transfer, disciplinary detention, or monetary restitution).
- Unit team members make up the Unit Discipline Committee in each housing unit. Wardens at institutions that do not operate under the Unit Management System should appoint one or more staff members to a committee authorized to hold initial hearings and impose minor sanctions.
- To ensure impartiality in any given case, neither the staff member who submitted the incident report, the staff member who conducted the investigation, nor any other staff member who was involved in any significant way in the incident may serve on a Unit Discipline Committee considering that case.
- The inmate should be given a written copy of charges to be considered by the Unit Discipline Committee within twenty-four hours of the time staff became aware of the incident and should be advised of his or her rights to appear at a hearing before the Unit Discipline Committee. A hearing before the Unit Discipline Committee normally should take place no more than three working days after staff become aware of the inmate's alleged involvement in the incident.
- The inmate is entitled to be present at the hearing, unless his or her presence would jeopardize institution security. Reasons for

excluding an inmate from a hearing must be documented in the official record. The inmate may waive the right to be present at the hearing by signing a statement to that effect, in which case the hearing will proceed in the inmate's absence. If the inmate waives the right to be present at the hearing, but refuses to sign the waiver, a staff member should document this in a signed memorandum, which is countersigned by a second staff member as witness. While entitled to be present at hearings, inmates are not entitled to be present for the deliberations of the Unit Discipline Committee.
- The inmate may make a statement at the hearing and may present documentary evidence. The Unit Discipline Committee should consider all evidence presented at the hearing (including the investigative report) and take one of the following actions:
  —Drop any charge in the moderate severity or low severity category, or resolve any charge in those categories informally. If a charge is resolved informally or dropped, the Unit Discipline Committee should order that the incident report be expunged from the inmate's central file, although a record of any informal resolution should be maintained by the institution's chief of security.
  —Find that the inmate did not commit the prohibited act cited in the incident report, in which case the matter is closed and the incident report is expunged from the inmate's central file.
  —Find that the inmate did commit the prohibited act cited in the incident report, in which case the Unit Discipline Committee may elect to impose minor sanctions.
  —Refer the case to the Disciplinary Hearing Officer, which is mandatory if the prohibited act falls within the greatest severity or high severity category (even if the Unit Discipline Committee finds in the inmate's favor) or if major sanctions may be appropriate.
- The Unit Discipline Committee's findings and disposition, as well as the names of members who participated in the hearing, are to be cited in the appropriate space in the incident report, and that portion of the incident report should be signed by the chairperson of the Unit Discipline Committee. The findings and disposition should be entered within one working day, and a written copy should be provided to the inmate by the close of business on that same day.
- A committee member who wishes to dissent from decisions of the Unit Discipline Committee may do so by preparing a written

memorandum for the record. Unless extenuating circumstances intervene, this memorandum of dissent must be submitted within three working days of the hearing and should be filed with other disciplinary records in the restricted portion of the inmate's central file. Unless obtainable under the Freedom of Information Act or other statutes, the memorandum of dissent should not be provided to the inmate.
- A record of the Unit Discipline Committee's proceedings (which need not be verbatim), along with any supporting documents, should be maintained in the inmate's central file. Adverse findings must also be entered into the inmate's chronological disciplinary record, one copy of which must be maintained in the inmate's central file and another copy of which may be maintained (for easier reference) by the chief of security.
- If the Unit Discipline Committee decides to refer the case to the disciplinary hearing officer, the committee should advise the inmate of his or her rights at any hearing before the disciplinary hearing officer. In particular, the inmate should be advised that he or she is entitled to be present at hearings before the disciplinary hearing officer, may have a staff representative appear on his or her behalf, and may have witnesses testify on his or her behalf. The inmate should be asked to provide the name of the staff member he or she wishes to have as staff representative, the names of witnesses, and a description of the testimony the witness may be expected to provide. The inmate's witnesses will be called only if their names are provided by the inmate in advance of the hearing. If an inmate elects not to appear at a hearing before the Disciplinary Hearing Officer, then he or she should sign a waiver indicating such. If an inmate declines to attend and refuses to sign a waiver, a staff member should prepare and sign a memorandum to that effect, which must be countersigned by a second staff member as witness. An inmate who does not attend a hearing before the discipline hearing officer still may have a staff representative appear on his or her behalf.
- An inmate placed in administrative segregation or other restricted status pending resolution of his or her case should remain there if the case is referred to the discipline hearing officer. The Unit Discipline Committee may not make final disposition of a case that has been referred to the discipline hearing officer.
- Time limits regarding completion of the investigation, and regarding hearings and findings of the Unit Discipline Committee, may be extended at the request of the inmate or of staff, provided

good cause is shown. Reasons for extensions must be documented in the record of the Unit Discipline Committee's proceedings. Extensions beyond five working days require the approval of the warden.

5. Disciplinary Hearing Officer

- Matters that cannot be disposed of by the Unit Discipline Committee are referred to the disciplinary hearing officer. The disciplinary hearing officer may be assigned to one or more institutions and is responsible for conducting independent, administrative, fact-finding hearings; issuing findings; and imposing appropriate sanctions.
- An alternate disciplinary hearing officer (for example, a fully trained and certified disciplinary hearing officer normally assigned to another institution) may be appointed to hold hearings in the event of an emergency, if the institution's regular disciplinary hearing officer is unable to hold hearings or if the regular disciplinary hearing officer has been the reporting officer, investigating officer, or a witness in any incident that might be referred to the Unit Discipline Committee.

The institution should provide the inmate with written notice at least twenty-four hours in advance of his or her appearance before the disciplinary hearing officer. The inmate should be advised of his or her rights before the disciplinary hearing officer (including the right to be present at the hearing, the right to make a statement and present evidence, the right to request the appearance of witnesses, and the right to have a staff representative) at the time the case is referred to the disciplinary hearing officer by the Unit Discipline Committee (see "Unit Discipline Committee," above). The inmate is entitled to be present throughout the hearing process, except during periods of deliberation, when institutional safety may be jeopardized, or, in some instances, when outside witnesses are called to testify. Hearings may be conducted in the inmate's absence if the inmate waives the right to appear.

- The disciplinary hearing officer should call those witnesses who have information directly relevant to the charges and who are reasonably available. This may include witnesses from outside the institution. There is no minimum or maximum number of witnesses that may be called, but the disciplinary hearing officer need not call repetitive witnesses. The reporting officer and other adverse witnesses need not be called if their knowledge of the incident is adequately summarized in the incident report and other investigative materials supplied to the disciplinary hearing officer. The disciplinary hearing officer should request submission of written statements from witnesses who are not available to testify, including outside witnesses.

The inmate may request the appearance of certain witnesses, but there must be an indication that the requested witness has information directly relevant to the charges. The disciplinary hearing officer may refuse to call numerous character witnesses that the inmate may request.

The disciplinary hearing officer may compel inmate witnesses to testify. Failure to appear at a hearing could warrant an incident report for failure to refuse an order given by a staff member. The disciplinary hearing officer may advise inmate witnesses that lying at the hearing would also warrant an incident report for providing false statements to a staff member. It may be necessary to take steps to protect an inmate witness who is called to appear at a hearing.

If, for security or other reasons, an inmate must be excluded from the hearing while a particular inmate is to testify, the reasons for the inmate's exclusion must be documented in the official record. Following the witness' appearance, the disciplinary hearing officer should inform the inmate of the substance of the witness's testimony, unless doing so would reveal confidential information or jeopardize institution security.

Witnesses requested by the inmate may be questioned by the inmate's staff representative. If the inmate has waived staff representation, the witnesses would be questioned by the disciplinary hearing officer.

- The disciplinary hearing officer should consider all evidence presented at the hearing and should take one of the following actions:
  — Refer the case back to the Unit Discipline Committee for further investigation or for disposition (if it is determined that the matter does not warrant the involvement of the discipline hearing officer).
  — Find that the inmate did not commit the prohibited act cited in the incident report.

—Find that the inmate did commit the prohibited act cited in the incident report, in which case the disciplinary hearing officer may impose any of the sanctions (whether major or minor) appropriate for the infractions in the particular severity category (see "Levels of Severity," above).

- If the inmate has escaped or is otherwise absent from custody, the disciplinary hearing officer may conduct a hearing in the inmate's absence. If sanctions were imposed during a hearing in absentia, and the inmate subsequently is returned to custody, then a rehearing should be conducted within sixty days of the inmate's return. All appropriate procedural requirements before the disciplinary hearing officer apply to this rehearing, except that written statements from witnesses may be used if those witnesses are not readily available to appear. At the conclusion of the rehearing, the disciplinary hearing officer may dismiss the charges, may affirm both the earlier decision and the sanctions imposed, or may affirm the earlier decision but reduce the sanctions. The disciplinary hearing officer, however, is prohibited from increasing the sanctions previously imposed.
- The disciplinary hearing officer should prepare a record of the proceedings, which need not be verbatim. The record should document that the inmate was advised of his or her rights and should also document the disciplinary hearing officer's findings, the evidence on which the findings were based, the sanctions imposed (if any), and the reasons why the specific sanctions were imposed.
- A written copy of the disciplinary hearing officer's findings should be provided to the inmate within ten days of the decision. A record of the hearing and the supporting documents should be placed in the inmate's central file. If the discipline hearing officer finds that the inmate did not commit a prohibited act, then the incident report and related documents should be purged from the inmate's central file.
- Adverse findings of the disciplinary hearing officer must be entered into the inmate's chronological disciplinary record, one copy of which must be maintained in the inmate's central file and another copy of which may be maintained (for easier reference) by the chief of security.
- The disciplinary hearing officer must advise the office responsible for sentence computation of any disallowance of good conduct time.

6. Confidential Information and Confidential Informants

- Great care must be taken to protect the identity of confidential informants. An inmate who is the subject of a disciplinary action may not be present when a confidential informant testifies at a hearing of the Unit Discipline Committee or the disciplinary hearing officer, and the substance of that testimony should not be communicated to the inmate. If confidential information is discussed in the incident report, the investigative report, the Unit Discipline Committee's report, the disciplinary hearing officer's report, or any other supporting documentation, that information must not be disclosed to the inmate. Inasmuch as the incident report and the reports of the Unit Discipline Committee and the disciplinary hearing officer customarily are provided to the inmate, it would be appropriate to include confidential information in separate attachments to those reports, which may be easily withheld. Such attachments should be maintained in the restricted portion of the inmate's central file.
- The inmate's staff representative need not know the identity of confidential informants. At the discretion of the disciplinary hearing officer, the substance of a confidential informant's testimony may be divulged to or challenged by the staff representative. The staff representative, however, may not challenge the reliability of the confidential informant.
- The Unit Discipline Committee and the disciplinary hearing officer are solely responsible for establishing the reliability of the confidential information they receive before using that information to support any finding. Conclusions as to the reliability of confidential information or confidential informants should be documented in a portion of the record not available to the inmate. Reliability may be determined by a record of past reliability on the part of the confidential informant or by any other reasonable factors. A staff member providing information obtained from a confidential informant should prepare a written statement on the frequency with which the confidential informant has provided information, the period of time during which the confidential informant has provided information, and the degree of accuracy of that information. The staff member is obligated to determine whether any basis exists for concluding that a confidential informant may be providing false information. Neither the Unit Discipline Committee nor the disciplinary hearing officer may con-

sider information that a confidential informant provides with the expectation of any kind of favor, from either the institution or the inmate who is the subject of the disciplinary action.
- Ordinarily, a decision that an inmate committed a prohibited act must be supported by more than one reliable confidential source. If there is only one source, then the confidential information must be corroborated by independently verified factual evidence linking the inmate charged to the prohibited act. Only under peculiar circumstances may uncorroborated information from a single informant be sufficient to support a finding that the inmate committed a prohibited act—for example, in the case of an unwitnessed assault, the statement of a seriously injured assault victim may be sufficient to support an adverse finding without corroborating evidence (unless the inmate charged in the assault is able to provide convincing evidence to the contrary).

7. Appeals

- Inmates may appeal disciplinary decisions through the regular inmate grievance process. The correctional agency may impose time limits on the filing of appeals or grievances. All relevant documentation should be forwarded to the reviewing official upon the filing of an appeal or grievance.
- The reviewing official for decisions made by the Unit Discipline Committee is the warden. The reviewing official for decisions made by the disciplinary hearing officer is the counsel for the correctional agency.
- The reviewing official may approve, modify, reverse, or remand any decision. If the reviewing authority modifies a sanction, it may only reduce it; the reviewing authority may not increase any valid sanction. If the reviewing authority remands a decision, it may do so with specific directions (such as ordering a rehearing).

The reviewing authority should consider whether the Unit Discipline Committee or the disciplinary hearing officer substantially complied with agency regulations on inmate discipline, based its decision on the facts (and on the greater weight of evidence in cases where there was conflicting evidence), and whether the appropriate sanction was imposed according to the severity level of the prohibited act and other relevant circumstances.

- If a case is remanded, neither the Unit Discipline Committee nor the disciplinary hearing officer may increase the original sanction. The Segregation Review Official may reduce the time an inmate must spend in disciplinary detention as a punishment for misconduct (see Part VII).

IX. USE OF NONLETHAL FORCE AND APPLICATION OF RESTRAINTS

Overview

The correctional agency may authorize staff to use nonlethal force, but only as a last alternative after all other reasonable efforts to resolve a problem situation have failed. When authorized, staff must use only that amount of force necessary to gain control over the inmate; to protect and ensure the safety of inmates, staff, and others; to prevent serious property damage; and to ensure institution security and good order.

In addition, staff may be authorized to apply physical restraints necessary to gain control over an inmate who appears to be dangerous because he or she is assaulting another individual, destroying government property, attempting to commit suicide, inflicting injury upon himself or herself, displaying signs of imminent violence, or is violent. The use of restraints in such emergency situations is considered a separate matter from using precautionary restraints (handcuffs, leg irons, etc.) in routine situations, such as transferring or escorting inmates from the prison to a courtroom or from one part of the institution to a cell in disciplinary detention, etc.

1. Types of Force

Two types of nonlethal force may be applied when circumstances warrant.

Immediate Use of Force

Staff members may use force or apply restraints immediately when an inmate's behavior constitutes an immediate, serious threat to the inmate, staff, or others or to property or institution security and good order. In an immediate use of force situation, staff members may respond with or without the presence or direction of a supervisor.

Calculated Use of Force or Application of Restraints

In cases where a problem situation must be resolved, but the inmate is in an area that can be isolated (such as a locked cell or a range) and there is no immediate or direct threat to the inmate or others, staff members need not act immediately to physically restrain the inmate. When there is time for the calculated use of force or the application of restraints, the staff must first determine if the situation can be resolved without resorting to force (see "Types of Restraints," below).

2. Principles Governing the Use of Force and Application of Restraints

- Staff members should attempt to gain the inmate's voluntary cooperation before using force. Force must never be used to punish an inmate.
- Staff members may use only that amount of force necessary to gain control over the inmate, in situations such as defending or protecting oneself or others, enforcing institution policies, preventing a crime, or apprehending an individual who has committed a crime.
- Restraints may be appropriate when used to prevent inmates from hurting themselves, staff, or others, or to prevent serious property damage. Restraints should be used only when other means of control have failed or are impractical. The warden must be notified immediately after restraints have been applied to render a decision on whether the use of restraints should continue.
- Restraints (such as handcuffs) may be applied on any inmate who continues to resist after staff members achieve physical control over him or her. Staff members may apply restraints to any inmate who is placed under control by the use of force team technique, which is a means of quickly and effectively gaining control over an inmate while minimizing injury to staff or inmate. If an inmate in a forcible restraint situation refuses to move to another area on his or her own, staff members may move the inmate bodily by lifting and carrying the inmate, provided the restraints themselves are not used for the lifting or carrying.

- The inmate should remain in restraints until he or she regains self-control and ceases to behave in a threatening or disruptive manner. Specific inmate acts necessitating the continued use of restraints must be documented, and all staff members observing actions justifying the continuation of restraints must submit separate memorandums recording their accounts of those actions.
- Except where the immediate use of restraints is required for control of the inmate, staff members must be specifically authorized by the warden to apply or continue restraints to an inmate who is in a cell in administrative segregation or disciplinary detention.
- Restraints must not be used as a method of punishing an inmate. Restraints must never be placed about an inmate's neck or face for any reason, or placed in any manner that restricts blood circulation or obstructs the inmate's airways. For example, staff members are prohibited from placing tape, cloth wraps, or bindings of any type around an inmate's mouth, nose, or neck; they may be held criminally liable for any injuries that may result from such misuse of restraints.
- Restraints must not be used in a manner that causes unnecessary physical pain or extreme discomfort. In any situation calling for the use of restraints, staff members in general (and the ranking officer on the scene in particular) must ensure that no undue pressure (such as on the inmate's chest, back, or neck) is being placed on an inmate's body while the restraints are being applied. Staff members may not hog-tie inmates or apply restraints improperly or too tightly.
- Hard restraints (such as steel handcuffs and leg irons) should be used only after soft restraints prove ineffective or have not been used effectively on a particular inmate in the past (again, the use of hard restraints in precautionary situations, as cited in the description above, is not affected by this provision). Medication, drugs, or injections may not be used to achieve restraint.
- All inmates placed in restraints must be monitored closely, until such time as the restraints are removed. Inmates may not be secured to a fixed object (such as a cell door or cell grille), except as specified in "Types of Restraints," below.
- All incidents involving the use of force and the application of restraints must be carefully documented. Reporting and investigating all use of force incidents are necessary to protect staff members from unfounded allegations and to eliminate the un-

warranted use of force. Whenever practicable, use of force incidents should be videotaped, and the videotapes should be reviewed (along with all other documentation) by headquarters staff.

3. Types of Restraints

- A progressive restraint policy employs the least restrictive method to control the inmate, as deemed necessary for the specific situation involved. The level of restrictiveness of the restraints may be based on the inmate's past or current behavior.
- Ambulatory restraints are soft and hard restraint equipment that permit the inmate enough freedom of movement to eat, drink, and attend to basic human needs without staff intervention. Whenever possible, ambulatory restraints should be used initially.
- If the inmate's behavior demands more restrictive or secure restraints, then staff members may resort to the most appropriate alternative to ambulatory restraints: hard restraints without waist chain or waist belt, hard restraints with waist chain or waist belt, four-point soft restraints with hard restraints for securing the inmate to the bed, and four-point hard restraints. In situations involving highly assaultive and aggressive inmates, progressive restraints may be used as intermediate measures in placing the inmate into, or removing an inmate from, four-point restraints.
- Four-point restraints are employed to restrain an otherwise uncontrollable inmate by the hands and feet to a bed or other stationary object. Soft restraints (made of vinyl or leather) should be used, unless such restraints have proven ineffective, in which case hard restraints should be used. Use of four-point restraints may only be used with the warden's direct authorization, and the warden may not delegate the responsibility for authorizing four-point restraints. The inmate should be placed in a face-down position on the bed (at least initially) to avoid possible complications resulting from vomiting. The bed should have a mattress, and a blanket or other covering should be provided. The inmate must be dressed as appropriate for the weather, and the inmate's clothing may only be removed for body searches or medical examinations. An inmate in four-point restraints should never be allowed to remain naked or without bed coverings placed over his or her body, unless medical personnel determine that it is necessary to do so. The inmate's position should be rotated periodically

to avert soreness or stiffness, and the inmate should be afforded the opportunity to use the toilet once every two hours, unless he or she is still offering active resistance.

Staff members must check an inmate in four-point restraints at least once every thirty minutes to ensure that the restraints are not hampering circulation and to assess the inmate's condition. In addition, the shift commander must check the inmate at least once every two hours, to determine whether the restraints had accomplished the intended effect of calming the inmate and whether the inmate may be released from four-point restraints. Medical staff should examine the inmate at the time he or she is placed in four-point restraints to ensure that the inmate is breathing properly and his or her circulation has not been hampered and to assess the inmate's physical and verbal responses. The medical staff should reexamine the inmate at least twice during every eight-hour shift. The appropriate executive-level official at the correctional agency's headquarters should monitor four-point restraint situations every eight hours, via telephone calls from the warden, and should be advised of the reasons for the use of four-point restraints. Each review and each opportunity for the inmate to use the toilet should be documented, using the unit log or a special form devised for documenting four-point restraint situations.

4. Confrontation Avoidance Procedures

- Confrontation avoidance techniques can defuse a potentially dangerous situation and avert the need to use force to control an inmate. Prior to any calculated use of force, the ranking security staff member (ordinarily the chief of security or the shift commander), a designated mental health professional, and others should confer and gather pertinent information about the inmate and the immediate situation. Based on their assessment of that information, they should identify staff who should attempt to obtain the inmate's voluntary cooperation and, using the knowledge they have gained about the inmate and the incident, determine if use of force is necessary. The careful documentation (including videotaping) of properly executed confrontation avoidance steps will demonstrate good faith on the part of staff members and show the inmate's willingness or unwillingness to comply with lawful orders.

- In calculated use of force situations, there usually would be time for the chief of security, the shift commander, the designated mental health professional, other pertinent staff (such as the inmate's unit manager, case manager, or counselor, or other officers or staff in the inmate's living or work area who may have a relationship with the inmate) to confer with each other and assess the situation. Such discussions may be conducted by telephone or in person, with the purpose being to gather relevant information about the inmate's medical and mental health history, and any recent incident reports or situations that may be contributing to the inmate's current state of mind. The assessment could include discussions with any other staff who are familiar with the inmate's background or present status. Information developed during the assessment might furnish insights into the cause of the inmate's agitation and assist in identifying staff who might have some rapport with the inmate and be more likely to succeed in reasoning with the inmate.

5. Use of Force

- Use of calculated force rather than use of immediate force is feasible in the majority of incidents that correctional practitioners are likely to encounter. Staff must use common sense and good correctional judgment in each situation to determine when there is time for the calculated use of force. The safety of persons involved is the paramount concern in all use of force situations.
- Immediate and unplanned use of force is required if an inmate is in the act of trying to inflict life-threatening injuries upon himself or herself, or is attacking a staff member or another inmate. Ordinarily, if those circumstances are not present, staff should employ the principles of calculated use of force.
- In calculated use of force situations, other staff (such as psychologists and counselors) can try to bring about resolutions in a nonconfrontational manner. Calculated use of force would be appropriate in cases where the inmate is verbalizing threats or brandishing weapons but poses no immediate danger.
- Calculated use of force episodes should be documented in writing (with a copy of such documentation being placed in the inmate's central file) and should be videotaped (with introductions of all staff participating in the confrontation avoidance group, as well as the actual confrontation avoidance process). The tape and docu-

mentation will be made part of the investigation package for the after action review process.
- If actual use of force is determined to be necessary, the use of force team technique should be used to control the inmate and apply restraints. The use of force team typically is made up of five or more staff members who are specially trained in use of force techniques. Each member of the team should wear protective gear (including helmets with face shields, jumpsuits, gloves, boots, and pads). The team should be briefed in advance by the supervisor in command of the team (usually a staff member of shift commander rank or higher) to identify each team member's specific responsibilities.

A typical approach to be followed under the use of force team concept is, first, to give the lead team member the assignment of pinning the disruptive inmate's body against the wall or floor. The second member then secures the upper left portion of the body, while the third member secures the upper right portion. The fourth member secures the lower left portion, and the fifth member secures the lower right portion. (Assigning members of the team to secure right, left, lower, and upper portions of the inmate's body *as they see them* is more preferable than assigning team members to secure specific body parts—such as the right leg or the left arm—because inmates are liable to tumble or thrash about when struggling with officers, thereby making it difficult to distinguish instantly between the right and left arm, or creating confusion if the staff member assigned to the left arm collides with the staff member assigned to the right leg as the inmate shifts position).

Once the inmate has been subdued, the third member of the team should be responsible for applying handcuffs (with the inmate's wrists positioned behind the back); the fifth member of the team should be responsible for applying leg irons.

Professionalism and discipline during any use of force exercise are of utmost importance. There should be no unnecessary talking or removing of equipment by team members during the entire use of force procedure. Team members should, as much as possible, protect the inmate's head and neck from injury during the exercise.

The supervisor in charge of the team should be positioned to observe the entire use of force incident but should not participate actively in subduing the inmate unless it is necessary to prevent staff or inmate injury.

- Other means for gaining control over an inmate may include the use of chemical agents, pepper mace, or nonlethal weapons (such as stun guns). Staff members may use only chemical agents or nonlethal weapons against individual inmates when the inmate is armed or barricaded, or cannot be approached without danger to themselves or others, and it is determined that a delay in bringing the situation under control would constitute a serious hazard to the inmate or others or would lead to a major disturbance or serious property damage.

Except where circumstances dictate immediate use, medical staff should be consulted before chemical agents or nonlethal weapons are used. Chemical agents in aerosol dispensers should not be discharged at a range closer than four feet from the inmate and should be directed in short bursts at the lower part of the inmate's face (avoiding the eyes). Pepper mace should not be discharged at a range closer than four feet from the inmate and should be directed in short bursts at the inmate's entire face (including the eyes). After being subdued, the inmate should wash all areas affected by a chemical agent with soap and water and should be examined by medical staff immediately and thereafter once a day for a minimum of two weeks, to detect any possible side effects resulting from the chemical agents.

- Use of force incidents must be documented and reviewed to determine if the use of force was warranted and carried out properly. The warden, deputy warden, chief of security, and chief medical officer should conduct an after-action review of the incident within one working day, and the warden should provide all documentation to the appropriate executive-level official at the correctional agency's headquarters within two working days, along with a report on the findings of the after-action review.

6. Use of Force Safeguards

In addition to wearing protective gear, use of force teams should observe several precautions before, during, and after calculated use of force incidents:

- Team members should receive instruction concerning communicable diseases during annual refresher training.

- Staff with skin diseases or skin injuries should not be permitted to participate in a calculated use of force action.
- Special precautions should be taken when using force on an aggressive inmate with open cuts, sores, or lesions, including use of all prescribed protective gear, a full body shield, and ample padding, and such inmates should be approached with great care.
- The chief medical officer should be consulted before attempting to use force against an inmate who is pregnant, or before attempting to use force on other special needs inmates, such as those with physical handicaps or suffering from mental illnesses.
- In cases where a pregnant inmate must be restrained, staff members should take every precaution to ensure that the fetus is not harmed. These precautions should be prescribed by institution medical staff, who should provide direction on the manner in which a pregnant inmate may be restrained and whether the inmate should be restrained at the institution hospital or at a medical facility in the community. In many cases medical staff members should be present during the application of the restraints.
- If there is spillage of blood or other body fluids during a use of force incident, the designated investigating officer or the shift commander must determine if there is a need to preserve any of it as evidence; in some cases, this decision may be made only with the authorization of the coroner or criminal investigator. Following this determination, and the collection of evidence as appropriate, staff or inmates wearing protective gloves should sanitize the cell walls, floors, etc., with appropriate disinfectant, and in accordance with standard sanitation procedures. In addition, any clothing or equipment that has been contaminated with body fluids should be disinfected or destroyed, as appropriate.

7. Custody Control Belt

- Remotely activated custody control belts, which can emit a nonlethal electrical shock in order to momentarily disable an inmate, may be used in conjunction with other restraints under certain clearly defined circumstances.
- Custody control belts may be used only on maximum-security inmates who are under escort.
- Use of a custody control belt may be authorized only by the warden of a high-, maximum-, or administrative-security institu-

tion or by his or her designee. The authorizing official must first determine for each specific case that conventional restraints are insufficient and that the inmate requires the additional security afforded by a custody control belt; further, the authorizing official must determine that the inmate's medical condition does not preclude use of a custody control belt. The justification for placing a custody control belt on an inmate must be documented in detail on the Escorted Trip Authorization Form.
- A physician, physician assistant, or nurse practitioner must review any request to use a custody control belt, to determine if any medical conditions exist that preclude use of the belt. Disqualifying medical conditions include (but are not limited to) pregnancy, heart disease, multiple sclerosis, muscular dystrophy, and epilepsy.
- Before placing the custody control belt on the inmate, staff must advise the inmate that the custody control belt will be placed on him or her, explain the circumstances under which the custody control belt might be activated, and indicate the physical effects that would occur if the custody control belt is activated. A notification form on this matter should be prepared for the purpose of informing the inmate, and this form should be read by or to the inmate. The inmate should sign the form. If he or she refuses to do so, staff members should document this on the form.
- All staff participating in an escort that involves the use of a custody control belt should be fully trained in technical aspects of its use and in agency policies governing its use.
- The officer in charge of an escort detail that involves the use of a custody control belt must be an employee at the rank of shift commander or higher. Only the officer in charge of the detail may carry and use the activating device for the custody control belt.
- The custody control belt may be activated only to prevent escapes or to prevent the loss of life or grievous bodily harm. In the event of an attempted escape, verbal orders to halt must be given to the inmate wearing the belt; if the inmate fails to halt immediately, the belt may then be activated. If the inmate has escaped and is out of view of the escorting officer, however, a verbal warning to halt is not necessary. Additionally, verbal orders are not required if the escorting officer reasonably believes that there is an imminent danger of death or grievous bodily harm due to the inmate's violent behavior.
- If the custody control belt is activated while being worn by an inmate, the officer in charge must notify the warden immediately,

submit a written report explaining the circumstances under which the belt was activated, complete appropriate sections of the forms used for documenting escorted trips and use of custody control belts, and submit a separate report documenting the use of force. The officer in charge should complete all such documentation and turn it over to the warden before the conclusion of his or her shift. The warden should then submit a full written report of the incident within one working day to the appropriate executive-level official at the correctional agency's headquarters.
- The inmate should undergo a medical examination as soon as possible following activation of the custody control belt, and the examination must be documented in the inmate's medical file. Any injuries, bruises, or marks on the inmate's body should be photographed or videotaped. If it is not possible for institution medical staff to carry out the examination—for example, if the custody control belt is activated during an escorted trip some distance away from the institution—then the escorting staff should take the inmate to a local community medical facility for the medical examination and any necessary treatment.

X. USE OF FIREARMS

Overview

Government-owned firearms may be issued to staff in certain clearly defined situations. Firearms may be discharged only in rare circumstances, and as a last resort. Government-owned firearms may only be carried and used by staff in the performance of their official duties. Staff are not permitted to carry or use any firearms other than government-owned firearms that have been issued to them in the performance of their duties.

1. Carrying of Firearms

- When authorized by the warden, staff members may be permitted to carry firearms when transporting inmates, when participating in escape hunts, or when assigned to security posts (such as towers) that require firearms as standard equipment. Post orders, riot plans, and escape plans should be written to include authorization and instructions for the carrying and discharge of firearms.

Staff members who escort inmates by commercial air carrier may also carry firearms, in accordance with Federal Aviation Regulations (Part 108.11, "Carriage of Weapons"). Staff members must comply with all provisions of that regulation, which authorizes law enforcement officials to carry arms in the performance of their duties, including completion of the required form before embarking upon the trip.

2. Discharge of Firearms

- Firearms may be discharged *as a last resort* to prevent an escape or attempted escape from a correctional facility. In such circumstances, verbal orders to halt must be given first. If the orders are not obeyed, a warning shot should be fired, but only if the staff member preparing to fire first determines that there is no apparent danger of injury to any innocent person. If the inmate or any person assisting the inmate continues his or her activities despite both the verbal orders to halt and the warning shot, firearms may be used to disable the potential escapee or person assisting in the escape. Verbal commands and warning shots, however, are not required if the escapee or person assisting the escapee initiates action to cause physical harm—in which case, staff members should use only the minimal force necessary to gain control of the inmate or the person providing assistance.

Firearms may not be used if the staff member recognizes the escapee as an inmate who was sentenced under juvenile statutes. Neither are firearms used to prevent escapes from minimum-security level institutions, unless the warden specifically determines that such use is necessary to protect the public and to prevent the loss of life or grievous bodily harm.

- Firearms may also be discharged when staff are escorting inmates outside the institution, under the same conditions and circumstances governing the discharge of firearms in escape situations. Heightened caution when using firearms in such situations is necessary, in order to avoid endangering the public. Firearms should not be discharged if doing so could endanger members of the public.
- Firearms may be discharged when there is a reasonable belief that the actions of another person are likely to result in the loss of life

or grievous bodily harm to staff, inmates, or others. Staff members should exercise sound judgment in making a decision about whether to fire a shot intended to disable, to fire a warning shot, or to issue a verbal warning.
- In hostage situations that have been contained, and where attempts to negotiate have begun, the discharge of firearms is under the immediate control of the warden or other official (such as an assistant director or regional director) who has assumed command of the situation. Only the warden or other official in command may order the discharge of firearms to resolve a hostage situation. Warning shots may not be fired in an attempt to resolve hostage situations or to create a diversion.
- Firearms may be discharged to prevent damage to or destruction of government property, in cases where the loss of this property could contribute directly to an escape or attempted escape, grievous bodily harm, or death. This would include firing on persons attempting to damage or disable a fire truck during a fire within an institution or inmates trying to break into a building where weapons or other security items are stored. Firearms should be used as a last resort, and in the same manner as for escapes (that is, a disabling shot may be fired following unsuccessful verbal orders to stop and a warning shot).
- Immediate medical attention must be provided to anyone injured during an incident involving the discharge of firearms.
- Any staff member who discharges a firearm while on duty must submit a written report to the warden before the completion of his or her shift. The warden should provide the appropriate executive-level official at the correctional agency's headquarters with a full report on any incidents involving the discharge of a firearm and should attach medical reports as appropriate.

3. Additional Considerations and Safeguards

- Staff assigned to carry firearms, except those carrying firearms at posts within the institution, should also be issued official badges to establish their authorization to carry firearms.
- Comprehensive firearms training should be mandatory for all new employees as part of the basic training program at the correctional agency's staff training academy. Staff should complete an approved firearms refresher training course at their home institutions once a year.

- No employees may be assigned duties requiring the carrying of firearms unless they have successfully completed the firearms training course. Chaplains, physicians, dentists, and others specifically exempted by the commissioner or director of the correctional agency should not be required to complete firearms training, but they may receive such training at their own request.

XI. ESCORTED TRIPS

Overview

At the discretion of designated institution staff, inmates may be granted the privilege of traveling outside the institution for specific purposes, while under staff escort. Escorted trips may be approved for such purposes as obtaining emergency or nonemergency medical treatment, visiting a critically ill member of the inmate's immediate family, or participating in program or work-related functions (such as certain educational activities). Such escorted trips may occur unexpectedly (as would be the case in a medical emergency), but, whenever possible, they should be planned in advance.

1. Escorted Medical Trips

- Escorted medical trips are intended to enable an inmate to receive medical treatment that would not be available within the institution. There are two types of escorted medical trips: emergency and nonemergency. Emergency medical trips may be authorized if an inmate experiences an unexpected, life-threatening medical situation that requires immediate medical treatment not available at the institution. The required treatment may be on either an in-patient or out-patient basis. In such cases, if the inmate's custody designation cannot be readily ascertained, then, for the purposes of the escorted medical trip, the inmate should be considered to be designated as "in" custody. Nonemergency medical trips may be authorized for a planned medical procedure that cannot be provided at the institution, and may be on either an in-patient or out-patient basis.
- Except in emergency situations, medical staff should complete an Escorted Trip Authorization Form in advance of any prospective trip, indicating the reasons for the trip. The chief medical officer

is ultimately responsible for determining whether an escorted medical trip is appropriate. Both the form and the inmate's central file should then be circulated to the case management staff for screening and clearance, to the institution's investigative supervisor for intelligence purposes, to the chief of security for assignment of escort staff and a determination of the required restraints, and to the appropriate unit manager and deputy warden for review and recommendation. The deputy commissioner in charge of medical services, or other appropriate official at headquarters, must be notified of all escorted medical trips.
- Ordinarily, reviews of requests for nonemergency medical trips should take place during regular working hours, in which case the warden should be the authorizing official. If for some reason a request must be reviewed during nonworking hours, then the duty officer or shift commander may serve as the authorizing official. In such cases, however, the warden must be notified as soon as possible if any medical trips are authorized in his or her absence.
- Inasmuch as time may be critical in an emergency medical situation, different procedures may be followed when authorizing an emergency medical trip—particularly if the emergency occurs during nonworking hours. If the warden is not available, then authorization may be given by the duty officer or the shift commander. Moreover, it is permissible for authorization to be given verbally, even before any paperwork is completed. In an emergency situation, it is permissible to complete the appropriate paperwork *after* the inmate has been escorted to an outside medical facility, rather than completing it beforehand.
- If an escorted trip for medical purposes is expected to result in an inmate being admitted to a community medical facility for in-patient treatment, the chief of security should arrange for custodial coverage with a predetermined contract guard service. (Contract guards may not be used for maximum-security inmates; for inmates in pretrial status, the appropriate local law enforcement agency should provide custodial coverage.)

The chief of security should develop post orders and logbook procedures that must be followed by correctional officers and contract guard services providing custodial coverage for an inmate receiving in-patient medical care, and the designated officers or guards must sign a statement indicating their awareness of the established procedures. The post orders and logbook procedures must be maintained by the escort officers for the duration of the inmate's community medical placement and must be returned to the institution thereafter. Unit staff should complete two sets of transfer orders—one to transfer the inmate to the medical facility and the second to return the inmate to the institution. The records office should produce sufficient copies of the transfer receipt to document the chain-of-custody when the inmate in placed in the custody of a contract guard service and when the inmate is returned to the institution.

2. Escorted Nonmedical Trips

- There are two types of escorted trips that would not involve medical treatment: escorted emergency/nonmedical trips and nonemergency/nonmedical trips. Escorted emergency/nonmedical trips may be approved to allow inmates to make a bedside visit to a critically ill member of their immediate family or to attend the funeral of a member of their immediate family. Parents, grandparents, stepparents, foster parents, brothers, sisters, spouses, children, and grandchildren would be considered immediate family under most circumstances. Escorted nonemergency/nonmedical trips may be approved for such purposes as enabling inmates to participate in program-related functions, including educational, work-related, or religious activities.
- If an inmate requests an escorted emergency/nonmedical trip, unit staff members should investigate the matter to determine if such a trip would be merited. They should use all available information and should contact attending physicians, hospital staff, funeral home staff, family members, probation officers, or others who might be able to furnish information on the validity of the inmate's request.
- If it is determined that an escorted emergency/nonmedical trip is appropriate, unit staff should complete an Escorted Trip Authorization Form and circulate the form and the inmate's central file to case management staff for screening and clearance, to the chief of security for assignment of escort staff and a determination of required restraints, and to the unit manager and the deputy warden for review and recommendation. Ordinarily, the warden would be the authorizing official, although, after regular business hours, the duty officer may be the authorizing official.
- The government should assume the salary expenses of the escort staff for the first eight hours of each day. All other expenses,

including transportation costs, must be assumed by the inmate, the inmate's family, or another appropriate source as approved by the warden. Unit staff, in consultation with the business office, should determine the portion of trip costs to be borne by the inmate; the inmate should complete a commissary withdrawal form—payable to the correctional agency—in that amount. If necessary funds are sent by outside sources, they must be deposited to the inmate's commissary account prior to the trip. If trip expenses turn out to be less than anticipated, the funds withdrawn that are in excess of the actual trip expenses must be returned to the inmate's commissary account.
- Escorted nonemergency/nonmedical trips may be considered for inmates who have been at the institution for at least ninety days, who are considered eligible for less secure housing, and who are considered eligible for work details under minimal supervision outside the institution. Ordinarily, escorted nonemergency/nonmedical trips would be available only to inmates with "out" or "community" custody.
- The department making the request (for example, the Education Department, or Chaplaincy Services) should complete the Escorted Trip Authorization Form and circulate it in the same manner as for escorted emergency/nonmedical trips (see above). Only the warden or acting warden may authorize an escorted nonemergency/nonmedical trip.

3. Inmates Requiring a High Level of Control and Supervision

- Except in medical emergencies, the special monitoring branch at the correctional agency's headquarters must give authorization before any inmate who is a protected or special witness may receive an escorted trip. In emergency medical situations, the inmate should be transported to the nearest community medical facility, in accordance with local procedures. In such instances, the special monitoring branch should be notified immediately by telephone, or, if the medical emergency occurs after regular business hours, the duty officer at the correctional agency's headquarters should be notified immediately.
- Except in medical emergencies, the appropriate executive-level official at the correctional agency's headquarters (such as the assistant director or assistant commissioner in charge of security

or correctional operations) must give authorization before any maximum or "in" custody inmate may receive an escorted trip. In medical emergencies, the inmate should be transported to the nearest community medical facility, in accordance with local procedures. In such instances, the assistant director or assistant commissioner should be notified immediately, by telephone, or, if the medical emergency occurs after regular business hours, the duty officer at the correctional agency's headquarters should be notified immediately.
- In determining if a maximum or "in" custody inmate may receive an escorted trip for other than emergency medical reasons, the assistant director or assistant commissioner should consider all relevant information, including the inmate's sentence, time in custody, and adjustment, as well as the nature of the request, and should maintain a written record citing reasons for any decisions.

4. Escort Procedures

- Except in medical emergencies, the inmate must agree in writing to all conditions of the escorted trip. Disciplinary action may be taken against an inmate who fails to comply with any of the conditions of an escorted trip.
- Inmates under escort must be within the constant and immediate visual supervision of escorting staff at all times.
- Restraints may be applied to an inmate going on an escorted trip, after the purpose of the escorted trip and the degree of supervision required by the inmate have been considered by the appropriate officials. Throughout all nonmedical escorted trips, including bedside visits and attendance at funerals, staff are required to use the minimum restraints as required under agency policy. An inmate who absconds from an escorted trip should be considered an escapee.
- Escorting staff are selected by the chief of security, in consultation with the hospital administrator, the unit manager, or others, as appropriate. The chief of security should designate on the approval form the specific staff member (ordinarily, the ranking member of the escort team) who will serve as officer-in-charge and have decision-making authority and responsibility throughout the escorted trip.
- Depending on the inmate's custody level and other conditions imposed by the warden, escorting staff may carry weapons. Re-

quirements regarding staffing, weapons, and restraints in any given escort situation remain in effect throughout the trip, including the duration of an inmate's inpatient status at a medical facility (unless exceptions are authorized by the deputy director or deputy commissioner).
- During emergency medical situations, the warden, after consulting with the chief of security and the hospital administrator, may consider exceptions to existing policies on restraints. For example, certain restraints may not be appropriate for pregnant inmates or inmates with broken limbs. In cases where it is inappropriate to apply restraints that otherwise would be mandatory, the warden should increase the number of escorting officers or consider authorizing escorting officers to carry weapons. All such deviations, along with the reasons for them, must be documented on the Escorted Trip Authorization Form.
- Specific custody requirements for the various custody levels appear in Exhibit 2–4. At least one staff member of the same sex as the inmate must be assigned to each escort team. Except in emergency situations, privately owned vehicles should not be used for escorted trips.

XII. VOLUNTARY SURRENDERS AND UNESCORTED TRANSFERS

Overview

Under appropriate circumstances, individuals who have been committed to a specific correctional facility, or who are being transferred from one correctional facility to another or to a halfway house, may be permitted to travel to their designated destinations without escort. Such voluntary surrenders (also known as unescorted initial commitments) and unescorted transfers (also known as furlough transfers) can expedite prisoner movement at a substantial savings to the government.

1. Voluntary Surrenders

- If the court recommends or orders a voluntary surrender commitment, a representative of the court should notify the correctional agency's community programs manager.

Exhibit 2–4 Custody Requirements for Various Custody Levels

Maximum Custody

- Contract guard services may not be used.
- A minimum of three staff escorts, including one at the rank of shift commander or above, must be used.
- In addition to three escorts with the inmate, there should be two staff members in a backup car (i.e., chase vehicle).
- At least one staff escort should be armed, and staff in the chase vehicle should be armed.
- Handcuffs with a handcuff cover, martin chains, and leg irons should be used at all times.*
- Staff should wear protective vests with a minimum threat level III classification.

"In" Custody

- Contract guard services may be used for "in" custody inmates who are at minimum- or low-security levels, but may not be used for those at medium- or high-security levels.
- A minimum of two staff escorts must be used.
- Escorting staff may be armed, at the warden's discretion.
- Handcuffs with martin chains should be used at all times, and other restraints may be used at the discretion of the escorting officers.*

"Out" Custody

- Contract guard services may be used.
- One staff member may escort a maximum of five "out" custody inmates, although more may be assigned at the discretion of the chief of security or the warden.
- No weapons are required, but they may be authorized by the warden.
- Restraint equipment may be used at the discretion of the escorting officers, or at the discretion of the warden.

Community Custody

- Contract guard services may be used.
- One staff member may escort a maximum of five community custody inmates, although more may be assigned at the discretion of the chief of security or the warden.
- No weapons are required.
- No restraints are required.

*Unless exceptions for medical or other reasons are appropriate, as noted in the policy on restraints.

- Upon notification that a voluntary surrender has been recommended or ordered, the community programs manager should request that the office at the correctional agency's headquarters in charge of inmate designations select an appropriate facility for service of the sentence. Once the designation has been made, the designated facility, the court, the offender, and the agency confining the offender (if the offender is confined) should be notified of the designation, and the date and time that the offender has been ordered to surrender.
- By certified mail, the community program manager should forward to the designated institution the original and one certified copy of the Judgment and Commitment Order, an authorized Unescorted Commitment and Transfer Form (with fingerprints and recent photograph attached), a copy of the designation, a statement showing the history of prior confinement on the current charge, and a presentence investigation report (if available).
- The designated institution will execute the Judgment and Commitment Order upon receipt of the offender, and the original executed judgment will be returned to the court. Any problems or questions raised by the offender, including a request for a delay in reporting, must be referred to the court.
- If the offender fails to report as instructed, the designated institution should notify the court and the appropriate law enforcement agency (such as the state police) by telephone by the end of the working day, and follow up with written confirmation (a letter or telefax). A copy of the confirmation should also be sent to the security administrator at the correctional agency's headquarters. In addition, the designated institution should follow any requirements specified under the victim and witness notification policy.

2. Unescorted Transfer to Another Prison within the Correctional Agency

- If an inmate is transferred to another facility within the system, and is authorized for an unescorted transfer, the records manager at the sending institution should forward (via certified mail) the following items to the records manager at the receiving institution at least seven working days before the scheduled transfer: the original transfer order, a copy of the Judgment and Commitment

Order, the Authorized Unescorted Commitment and Transfer Form (with fingerprints and recent photograph attached), and a copy of the Furlough Application and Approval Form (which would include the inmate's travel schedule). The preparation and mailing of these materials should be documented in the inmate's central file.
- The records manager at the sending institution should notify the records manager at the receiving institution as soon as the inmate has departed, and, as part of the notification, verify the inmate's travel schedule. This notification should be documented in the central file.
- The records manager at the receiving institution should notify the records manager at the sending institution of the inmate's arrival. No later than one working day after receiving notification that the inmate has arrived at the receiving institution, the records manager at the sending institution should forward (via certified mail) all files to the records manager at the receiving institution (including the inmate's judgment and commitment file, central file, medical file, etc.).
- If the inmate fails to report as directed, the receiving facility should notify the court, the appropriate law enforcement agency (such as the state police), the security administrator at the correctional agency's headquarters, and the sending institution.
- The sending institution is responsible for updating appropriate records to indicate a change in the inmate's release status from unescorted transfer to escape, and for making the inmate's sentence computation inoperative as of the day following the inmate's failure to report. The sending institution is also responsible for preparing the incident report, conducting a disciplinary hearing in absentia, and following procedures in the policy on escapes (including notification to judges, prosecutors, probation and parole officers, other officials, victims, and witnesses).

3. Unescorted Transfer to a Halfway House

- Not less than two weeks before the transfer, unit staff at the sending institution should forward (by certified mail) the following items to the designated halfway house: Authorized Unescorted Commitment and Transfer Form (with fingerprints and recent

photograph attached), a copy of the Furlough Application and Approval Form (which should include the inmate's travel schedule), the original transfer order, and a receipt for halfway house policies and regulations and the signed subsistence agreement (if applicable).
- The records manager at the sending institution is responsible for ensuring that the inmate's sentence computation is updated and complete, with all good time entered before the inmate departs the institution. On the date of the transfer, the records manager at the sending institution should notify the community programs manager at headquarters of the inmate's departure and travel schedule, and document that notification in the inmate's central file.
- The records manager at the sending institution must confirm the inmate's arrival at the halfway house. Within one working day of the inmate's arrival at the halfway house, the records manager should forward (via certified mail) the following items to the community programs manager at headquarters: all required release paperwork, the Victim Witness Notification (if applicable), the Committed Fine Form and all related documentation, such as the presentence investigative report (if applicable), copies of the Conditions of Supervised Release Form (if applicable), and appropriate forms and other documentation concerning final good-time awards.
- If the inmate fails to report to the halfway house as directed, staff at the halfway house should notify the community programs manager within twenty-four hours. The community programs manager should notify the appropriate law enforcement agency (such as the state police), the security administrator at headquarters, and the records manager at the sending institution of the inmate's escape.
- The sending institution is responsible for updating the records to indicate the change in release status from unescorted transfer to escape, as of the day the inmate failed to report, and for making the inmate's sentence computation inoperative as of the day following the inmate's failure to report. Staff at the sending institution should also prepare an incident report, conduct a disciplinary hearing in absentia, and observe applicable procedures in the policy on escapes (including notification of judges, prosecuting attorneys, probation and parole officers, other officials, victims, and witnesses).

XIII. SECURITY CONSIDERATIONS RELATING TO INMATE ACCESS TO COMPUTERS

Overview

Correctional agencies and institutions, like other government agencies, private sector corporations, and other organizations, should have sound computer security procedures in place. Such procedures would include password and personal identification controls, virus scanning, backing up data on a routine basis, secure storage of backup disks, ensuring the physical security of computers (to protect against both unauthorized use and accidental damage), background investigations on staff before clearing them for access to sensitive data, and verification that copyrighted software has been properly licensed. To help ensure that these and other procedures are being observed, and that computer security problems are reported and rectified, the correctional agency should appoint a computer security officer to establish and direct the system's computer security program. The wardens at each institution should establish a Computer Security Committee, chaired by an institution computer security officer, to oversee local computer security matters.

Of particular importance in a correctional environment is the establishment of systems and procedures to prevent inmates from misusing computers (including using computers to gain access to restricted information, to compromise institution security, or for fraudulent or other criminal purposes). Suggested systems and procedures are outlined below.

1. Allowable Activities

- Inmates may use computers for certain tasks, but only within very strict limitations. Inmates who are assigned to perform clerical duties as part of their required work programs may perform routine data entry and retrieval using specific, approved, commercial software, provided the data are not sensitive or restricted, and provided the inmates do not use the approved software to write their own programs.
- Inmates may have access to computers that are used in occupational training and education programs. For example, inmates may receive introductory training on how to operate personal computer (PC) hardware, operating systems, word processing,

drafting, spreadsheet design, and business-oriented application software. Training or access to publications providing instruction in programming techniques, programming languages, or database commands should be prohibited. In addition, inmates may participate in education classes (such as basic literacy, English as a Second Language, etc.) that use computers as teaching aids.
- Electric typewriters, nonprogrammable word processors, electronic calculators, electronic dictionaries, and other similar electronic devices ordinarily would not be considered computer systems, and therefore may be used by inmates unless it can be determined that they threaten the orderly management of the institution. (Diskettes from nonprogrammable word processors, however, do need to be controlled in accordance with provisions outlined in "Special Precautions," below). Inmates may have access to task-specific PCs, kiosks, CD-ROMs, etc., where there is no possibility for inmates to reprogram memory, manipulate files, or change data.
- Inmates should be allowed access only to hardware and software needed for the specific task to be performed. Access to DOS or other operating systems may be authorized if executable files and programs not necessary for normal operation are removed or restricted.

2. Prohibited Activities

- Inmates should never have access to data that are not public information. Inmates should never have access to a system that possesses or stores sensitive data or that is labeled "Staff Only."
- Inmates should not be allowed to develop software; this would include software using traditional programming languages such as BASIC, C, and Pascal, application languages such as dBase and Clipper, and macro and formula languages of the various word processing and spreadsheet packages. Although inmates may use macros, they must not be permitted to create them. Neither should inmates have access to compilers (Turbo Pascal, Quick C, etc.) and interpreters (BASIC, BASICA, GWBASIC, etc.) that enable programs to be converted to a form the PC can execute.
- Under no circumstances should an inmate be permitted to use a terminal, workstation, or PC that has access to the institution's automated inmate information system. Nor may inmates have access to or possess output from the inmate information system,

unless the output has been screened for sensitive data and approved in writing by a knowledgeable staff member. Inmates are prohibited from access to a LAN unless the security programs section has granted its approval in writing. (This prohibition does not include previously approved software systems, with the stipulation that any such networks may not be connected to systems designed for staff use only or for the processing of sensitive information.)

- Inmates must not use terminals, workstations, or PCs connected to a modem or any device permitting communication with another device or computer external to the institution, except limited-distance and synchronous modems attached to dedicated or X.25 address-controlled circuits.
- Inmates should be prohibited from using any communications or utility software that can permanently erase, modify, or hide files. Inmates may use applications within communications software to perform assignments, but access to dial-up modems and phone lines should be prohibited. Inmates should not have access to multifunction applications that include software for dial-up modems, unless communications modules have been disabled or removed or all potential for inmate access to modems and phone lines is eliminated.
- Under no circumstances should an inmate be assigned to repair PCs or peripherals, including add-on boards, printers, monitors, or the system unit itself.
- Nonsensitive computer printouts may be authorized for inmate possession, but inmates must not be permitted to use computer equipment for personal needs (such as personal correspondence) without the warden's permission (on a case-by-case basis). Printing of materials not required for education courses or materials not specifically approved in advance should be considered unauthorized use of equipment and may subject the inmate to disciplinary action.

3. Special Precautions

- Clear distinctions should be made between computer systems and printers for the exclusive use of staff and PCs that may be used by inmates as well as staff. Labels should be prominently affixed to institution PCs, terminals, workstations, and printers, indicating either "Staff Only" (in red letters), or "Inmate Access" (in blue

letters). Any equipment with access to the automated inmate information system or other sensitive information should be labeled "Staff Only."
- Control of computer disks is critically important. All removable media (including diskettes, tapes, etc.) should be under strict staff control. Inmates should be under surveillance when possessing such media (read-only media, however, are excluded from this requirement). When inmate use of removable media is required (preparing backups, workstation booting, etc.), that use should be under the direct and constant supervision of staff. An inmate may be allowed personal possession of removable media to perform assigned duties or participate in occupational training and education, provided a practical accountability method, vetted by the computer security officer, has been implemented.
- Inmates should not be permitted to receive disks, tapes, or other electronic media from outside sources (except for authorized audiotapes), and should not be permitted to mail such media out of the institution. Any media (hard copy or electronic) that may provide technical information detrimental to security or the orderly operation of institution computers should be reviewed prior to inmate possession and may be confiscated if necessary.
- Since the use of PCs that depend on diskettes as their only storage medium increases the potential for abuse, new PCs should contain hard drives and existing PCs should be upgraded (when possible) by installing hard drives. (Computers used for network workstations are exempt from this requirement.)
- PC data security software should be installed to protect all PCs that inmates use. (Systems used strictly for educational purposes, read-only devices, LANs approved for inmate access, workstations without hard drives, and task-specific, read-only computers may be exempt from this requirement.)
- Only staff members may install monitoring software, in order to prevent inmate access to any part of the operating system.
- Monitoring software on inmate-access PCs should be configured to prevent inmates from having access to any serial or parallel ports not necessary to accomplish assigned tasks.
- Where monitoring software installation is required, the maximum security configuration option and area permissions should be used. Functions or areas not needed to perform assigned duties should be restricted.
- Monitoring software on inmate-access computers should be configured to require each inmate and staff member to have his or her

own user ID and password and should limit access strictly to the directories, programs, data, and computer resources required to perform assigned and approved tasks. Inmates should never be permitted to use another person's ID or password. If use of another person's ID or password is necessary in order to perform maintenance on a computer, then the ID and password should be changed following completion of the maintenance.
- Inmates with extensive computer expertise or histories of computer fraud must never be assigned to work involving computer use or allowed access to data or applications (except as specifically authorized by policy, such as participating in occupational training programs). This prohibition would cover inmates with a significant level of computer education beyond the basic skill level; education or experience as a computer programmer; expertise in electronic communications technology; or a knowledge of computer hardware, firmware, or software sufficient to compromise the institution's security and orderly management, bypass security hardware or software, or prevent detection of any unauthorized activity. This prohibition would also cover inmates who have any history of using computers to carry out computer activities or whose misuse of computers at the institution has resulted in a disciplinary report with a finding of guilt.

CHAPTER 3

Inmate Entitlements

I. INMATE TELEPHONE REGULATIONS

Overview

While inmates do not enjoy any constitutional rights to unrestricted access to telephones, it is often in the best interests of sound correctional management to permit inmates to enjoy limited access to telephones. Telephone privileges—ordinarily at the inmate's own expense—can help inmates maintain community and family ties, and thereby can be a valuable tool in the overall correctional process to better prepare inmates for eventual release. Telephone privileges may also contribute to good institutional morale. Further, under certain circumstances telephone access to attorneys can facilitate an inmate's right to effective counsel.

Nevertheless, telephone privileges must be limited, for reasons of public safety and institutional security. Inmate telephone calls (other than calls protected by attorney-client privilege) may be monitored and other restrictions imposed in order to detect and deter criminal activities, violations of institution regulations, or threats to institution security. Specific regulations on inmate financial responsibility for telephone calls should be in place to prevent abuses. Finally, telephone privileges may be curtailed as a disciplinary sanction or as a measure to ensure the security or good order of the institution.

1. Official Telephone Lists

- Each inmate may call individuals whose names and telephone numbers appear on a list that they have prepared and that has been approved by the institution. Ordinarily, the correctional agency should establish a maximum number of names that may appear on any inmate's list of approved telephone contacts, although the warden or deputy warden may authorize placement of additional names on the list in consideration of special circumstances (for example, if an inmate has a large family).

- Any inmate who chooses to have telephone privileges should prepare a proposed telephone list at time of admission and orientation. The inmate may submit telephone numbers for anyone they wish, including judges, elected officials, and members of the news media, provided that the inmate acknowledges that—to the best of his or her knowledge—the individuals on the list are agreeable to receiving calls and that all calls would be for purposes allowable under the regulations of the correctional agency and the institution. Attorneys may be included on the inmate's telephone list, with the understanding that calls would be subject to monitoring; arrangements for unmonitored calls to attorneys must be made separately (see "Inmate Telephone Calls to Attorney," below).

- After the inmate submits his or her proposed telephone list, unit staff should take approximately ten days to process the list. Ordinarily, immediate family members (parents, stepparents, foster parents, siblings, spouses, children, stepchildren, grandparents, or grandchildren) will be approved automatically for inclusion on the list, as will individuals already approved for the inmate's visitors list (see Part III). Staff should send notification letters to all other individuals on the list, indicating that the inmate has requested their names and numbers be placed on his or her telephone list and requesting the recipients to advise the institution, in writing, if they do not want to receive calls from the inmate—in which case, their names and telephone numbers will be removed from the inmate's telephone list. If two individuals sharing a telephone are both included on the inmate's telephone list, and those individuals disagree as to whether or not they wish to receive calls from the inmate, a final determination will be made based on the wishes, in writing, of the person in whose name the telephone is listed.

- In addition to withholding approval for including on the list the names and numbers of any individuals who have expressed an unwillingness to receive telephone calls from the inmate, the deputy warden is authorized to disapprove the inclusion of other individuals, such as:

     —victims or witnesses connected to the inmate's criminal conviction
     —inmates at other correctional facilities (as with inmate-to-inmate correspondence, inmate-to-inmate telephone calls require the approval of the wardens at both facilities)
     —current or former employees of the correctional agency, unless their inclusion on the list has been approved, in writing, by the warden
     —any individuals who appear to be included on the list for purposes of using telephone communications to facilitate an escape, introduce contraband, or engage in other criminal activities (information indicating such intentions may be obtained from outside law enforcement agencies)
     —any individuals who appear to be included on the list for purposes of using telephone communications to conduct business activities, in violation of the correctional agency's prohibition against inmates operating businesses
- Once a list has been approved, either in whole or in part, it would still be subject to revision. An individual approved for inclusion on the list may submit a written request to be removed from the list, in which case calls to that person will be suspended until the request can be verified; once verification has been obtained, the individual's name will be deleted. Names may be deleted from the list if it becomes apparent that the inmate is calling that individual for purposes that violate regulations or threaten the orderly management of the institution. Telephone sanctions (that could affect the telephone list) may be imposed following disciplinary hearings with a finding of guilt. And, once per quarter, the inmate may request telephone list changes—in which case, staff ordinarily should process the request within ten days.
- If a name is not approved for inclusion on the telephone list, or if a previously approved name is deleted from the list, the deputy warden should notify the inmate in writing within three working days. This notification may be delayed, however, for law enforcement purposes (for example, if the decision is based on suspicions of criminal activity by the inmate or the recipient of the call). Inmates may appeal adverse decisions regarding their telephone lists through regular inmate grievance procedures, and should be advised of this when they are informed of any such adverse decisions.

2. Procedures and Expenses

- Inmates ordinarily are responsible for all costs of making telephone calls, including fees for replacing lost telephone access codes (see below). Tolls, taxes, and other expenses of making calls should be deducted from the inmate's commissary account.
- The placement and duration of any telephone call is subject to the availability of funds in an inmate's commissary account. In addition, the warden may limit the maximum length of telephone calling based on the situation at the institution (such as the size of the inmate population and the demand for telephone use).
- While collect calls normally would be prohibited, the warden may authorize placement of collect calls under special circumstances. Collect calls might be permitted, for example, for new commitments and transfers, inmates in holdover status, inmates in protective custody, and inmates without funds. A typical provision of the telephone policy might permit new inmates to make a total of thirty minutes of telephone calls on a collect basis during the first thirty days at the institution.

A form should be developed that inmates without funds may use to request the privilege of making at least one collect call per month (with a maximum duration specified by institution policy); the inmate would submit the request form to the unit manager, who should render a decision within five working days. Permission to make collect calls should be in force for only thirty days; if the inmate remains without funds at that time and wishes to continue making collect calls, he or she would have to submit a new request. The commissary accounts of inmates requesting collect calling privileges must be monitored by the unit team to discern any pattern of depleting the account before making a request for collect calling privileges, and then replenishing it after collect calling privileges have been granted. If such actions persist over a period of six months, then it may be considered a pattern of abuse of telephone privileges that could warrant disciplinary action.

- Under compelling circumstances, the warden may authorize the use of government funds to pay for an inmate's telephone use. Examples of compelling circumstances would include situations where an inmate has lost contact with his or her family or when there is an emergency involving an inmate's family.

- Institutions with automated, debit-billing, inmate telephone systems should assign unique telephone access code numbers to each inmate. Telephone access code numbers should be considered items of value, and inmates who furnish their access code numbers to other inmates or accept another inmate's access code number may be subject to disciplinary action. Staff members should deliver telephone access code numbers to inmates in a confidential manner (for example, via regular institution mail, in a sealed envelope). Access code numbers that are lost or compromised must be reported immediately to staff, and those numbers should then be deactivated (and, if appropriate, replaced).
- Third-party billing and electronic transfer of calls to third parties should be prohibited. Except for collect calls or calls at government expense, as authorized by the warden, inmates may not place calls unless all expenses associated with the call can be directly and immediately deducted from the inmate's commissary expense. For example, calls to 1–900 numbers, 1–976 numbers, 1–800 numbers, and to credit card access numbers should be prohibited.
- Inmates may not use telephones to perpetuate frauds, engage in criminal activities, operate a business (even if the business enterprise is noncriminal in nature), or make threatening or harassing calls. Any violation of telephone regulations may be considered an abuse, subject to possible disciplinary action.

3. Monitoring of Inmate Telephone Calls

- Procedures and equipment should be in place to enable the institution to monitor conversations on any telephone on institution premises, in order to preserve the security and orderly management of the institution and to protect public safety.
- Staff must advise all inmates that their telephone conversations may be monitored. In addition, notices (in English and Spanish, and in other languages if deemed appropriate in light of the ethnic makeup on an institution's inmate population) should be placed at all monitored telephone locations, indicating that all conversations from that telephone are subject to monitoring and that use of the telephone constitutes consent to such monitoring. The notice should also state that inmates wishing to place unmonitored calls to their attorneys should contact a member of

their unit team. Staff may not monitor an inmate's properly placed call to an attorney (see "Inmate Telephone Calls to Attorneys," below).
- As part of the admission and orientation process, inmates should be advised of procedures for placing unmonitored telephone calls.

4. Inmate Telephone Calls to Attorneys

- Under certain conditions, inmates may make unmonitored calls to their attorneys of record. The correctional agency should ensure that inmates have several methods by which to maintain confidential contact with their attorney. Confidential inmate-attorney correspondence is covered under special mail requirements; private inmate-attorney visits are permitted; and inmates may be afforded the opportunity to make occasional unmonitored calls to their attorneys.
- Frequent confidential inmate-attorney calls should be allowed only when an inmate demonstrates that communication with his or her attorney by other means is not adequate (as could be the case, for example, when a trial date is imminent). If correspondence, visiting, or normal telephone use is not adequate for inmates to maintain necessary confidential contact with their attorneys, then the warden may not limit the frequency of unmonitored inmate-attorney telephone calls; otherwise, the warden may limit such calls to a certain number per month.
- Staff members must verify that the unmonitored calls they place on behalf of an inmate are, in fact, to the office of the inmate's attorney. Third-party calls are not authorized, calls should not be forwarded to others, no one other than legal staff should be involved in the calls, and attorneys or other legal staff receiving the calls should be advised that the calls are permitted exclusively for the purpose of discussing legal matters.

II. INMATE CORRESPONDENCE

Overview

Inmates should be encouraged to correspond with individuals outside the institution, if that correspondence is directed toward socially useful goals. The right of inmates to correspond with their attorneys, in confidence, must be preserved. Nonetheless, the institution may enforce certain restrictions on inmate correspondence, in order to preserve institutional safety and security and to protect public safety. All classes of correspondence may be inspected for contraband, some correspondence may be read by staff, certain types of mail may be confiscated or returned, inmates may be denied the privilege of corresponding with certain individuals, and limitations may be placed on the amount of postage an inmate may possess.

1. General Correspondence

- General correspondence is incoming or outgoing mail or packages *other than* special mail, legal mail, and publications (see below).
- There are two types of general correspondence: open general correspondence and restricted general correspondence.
- Open general correspondence does not limit an inmate's privilege of sending or receiving mail to a list of authorized correspondents. Open general correspondence may be given to those inmates who are able to accept the privilege in a responsible and mature manner.
- Restricted general correspondence limits an inmate's privilege of sending or receiving mail to a list of authorized correspondents. Inmates may be placed on restricted general correspondence based on misconduct or as a matter of classification. Determining factors may include the inmate's involvement in any effort to send or receive contraband through the mail or to circumvent correspondence regulations in any other ways (see below), or involvement in any attempts to secure noncontraband funds or items through the mail in an inappropriate or fraudulent manner. Other factors may include the security risk posed by the inmate, threats made by the inmate against any other person, or a criminal history involving violation of postal laws or any offense involving the mails.

A determination to place an inmate on restricted general correspondence should be made by the unit team or the unit classification team at the time of classification, reclassification, or disciplinary action. In cases where placement on restricted general correspondence is intended as a sanction for misconduct, staff must follow all procedures under inmate disciplinary regulations that are necessary before sanctions may be imposed. In other cases, the inmate should be afforded the opportunity to respond to the classification or reclassification in writing, or by appearing in person before the unit team or classification team. The final decision on whether or not to place an inmate on restricted general correspondence rests with the warden or deputy warden. Inmates placed on restricted general correspondence may appeal the decision under regular inmate grievance procedures (see Part IV).

Inmates on restricted general correspondence ordinarily may correspond with spouses (including common-law spouses), parents, and children (except those who have been involved in a previous violation of correspondence regulations or who would pose a threat to institutional security or public safety). The inmate may correspond, *for social purposes only*, with former business associates, unless the warden determines that such correspondence could threaten institutional security or lead to criminal activity. The inmate may request that other persons be placed on the list of authorized correspondents, subject to investigation, evaluation, and approval by the warden; with prior approval, the inmate may write to a proposed correspondent for the sole purpose of obtaining a release that would authorize an investigation. Finally, the warden may permit the inmate to have correspondence of a non-ongoing nature with persons who are not on the authorized list of correspondents if the inmate can demonstrate that such correspondence is necessary; authorization of such special purpose letters ordinarily would be the responsibility of the case manager.

- All *incoming* general correspondence may be opened and inspected by staff. This may be done as frequently as deemed necessary to maintain security or to monitor a particular problem or inmate.
- Ordinarily, the *outgoing* general correspondence being mailed by pretrial detainees, minimum-security inmates, and low-security inmates may be sealed by the inmate and sent out unopened and uninspected. Staff, however, are authorized to open any such outgoing general correspondence at any time if there is reason to believe that it would interfere with orderly management of the institution, it would be threatening to the recipient, or it would facilitate criminal activity. Such outgoing general correspondence may also be opened and inspected if the inmate is on restricted general correspondence, if the letter is addressed to another inmate, or if the envelope has an incomplete return address.

- *Outgoing* general correspondence being sent by medium- and maximum-security inmates may not be sealed by the inmate and may be read and inspected by staff.
- The primary objective of inspecting incoming or outgoing mail is to detect contraband. All incoming general correspondence should be inspected for contraband.
- The primary purpose of reading incoming or outgoing mail (usually on a random basis) is to detect escape plots, plans to commit illegal acts, plans to violate institution guidelines, or other concerns relating to public safety or institution security.
- The warden may reject correspondence that includes matter considered nonmailable under law or postal regulations; matter that depicts, describes, or encourages activities that may lead to physical violence or group disruption; information regarding escape plots, plans to commit illegal activities, or plans to violate the correctional agency's policies; information written in code; threats, extortion demands, obscenities, or gratuitous profanity; sexually explicit materials that could pose a threat to institution security or personal safety (for example, sexually explicit photographs in a magazine may not fall into this category, but sexually explicit personal photographs of a spouse, friend, relative, or acquaintance of an inmate could engender provocative comments, misunderstandings, or friction with other inmates and thereby could undermine orderly operations). Correspondence may also be rejected if it concerns attempts by an inmate to direct his or her business enterprises (this prohibition does not affect pretrial inmates who are still entitled to oversee their businesses; nor does it apply to efforts on the part of an inmate to protect property and funds that were legitimately his or hers at the time of commitment).
- Although inmate mail may typically be read on a random basis, institutions may wish to give closer scrutiny to the incoming and outgoing mail of certain inmates. For example, the correspondence of inmates who participated in unusually sophisticated criminal activities (especially those involving mail fraud), inmates considered escape risks, and inmates presenting management problems may be subject to intense monitoring. Such intense monitoring, however, must not interfere with the prompt handling of the mail. The warden should designate a staff member to supervise inmate correspondence and maintain a list of inmates whose mail is to receive added scrutiny.

- Staff must be sensitive to the fact that most information in inmate correspondence is of a private nature and must be handled in a discreet and professional manner. Staff must receive explicit instructions that the contents of an inmate's correspondence should never be revealed to any other person for any reason, unless there is a legitimate correctional concern relating to security, safety, the orderly running of the institution, or criminal activity.
- An inmate may be permitted to correspond with inmates in any other correctional facility, provided that they are relatives or are both parties in a specific legal action. Other correspondence between inmates may also be approved under exceptional circumstances, with consideration given to the inmates' security classifications and the nature of the inmates' relationship.

Correspondence between inmates at different facilities should be approved by the wardens at both institutions and should always be inspected (both at the institution where the correspondence was mailed and at the institution where it was received).

- The warden must give prior approval for an inmate to send or receive any package through the mail.

2. Special Mail

- Special mail is correspondence sent to elected government officials (including the president and vice president of the United States, governors, members of Congress or state legislatures, etc.), judges and other court officials, prosecuting attorneys, cabinet members, law enforcement officers, directors of correctional agencies, probation and parole officers, attorneys, and members of the news media.
- *Outgoing* special mail should be sealed and marked "Special Mail" by the inmate, and it may not be opened, read, or otherwise inspected by staff.
- Staff members should stamp the following message on the back of all envelopes containing an inmate's outgoing special mail: "The enclosed letter was processed through special mailing procedures for forwarding to you. The letter has been neither opened nor inspected. If the writer raises a question or problem over which this facility has jurisdiction, you may wish to return the letter or a copy of the letter for further information or clarification. If the

writer encloses correspondence for forwarding to another addressee, please return the enclosure to the above address." The stamp should also include the name and address of the institution (placed above the message) and space for the date to be entered.
- *Incoming* special mail may be opened only in the presence of the inmate and may be inspected only for physical contraband and to ensure that enclosures qualify as special mail. Special mail, however, may not be read by staff.
- Incoming special mail should carry a complete return address that identifies the sender and should be marked on the front of the envelope with the following notation: "Special Mail—Open only in the presence of the inmate," or words to that effect. Letters from judges or from members of Congress or state legislatures should be handled as special mail even if they do not carry that statement. Otherwise, incoming special mail that is not appropriately marked or that fails to identify the sender may be handled as general correspondence, subject to inspection and review by the staff.
- Suspected abuse of special mail privileges should be referred by the warden to the correctional agency's legal counsel.

3. Legal Mail

- Legal mail should be considered a form of special mail and generally should be handled in the same fashion as special mail (see "Special Mail," above). Ordinarily it would include correspondence between the inmate and his or her attorney, and between the inmate and the courts.
- Incoming legal mail should be marked on the front of the envelope with a notation such as "Special Mail—Open only in the presence of the inmate," "Legal Mail—Open only in the presence of the inmate," or "Attorney-Client Mail—Open only in the presence of the inmate." The inmate is responsible for advising any attorney that correspondence will be handled as legal mail only if the front of the envelope is marked is this way. In addition, the return address should include the name of the law firm and the name of the individual attorney. The return address on legal mail from an attorney's assistant or from a legal aid student or assistant must also carry the name of the attorney or the legal aid supervisor.
- Staff should mark each envelope containing incoming legal mail to show the date and time of receipt, the date and time the letter

was delivered to the inmate and opened in his or her presence, and the name of the staff member who delivered the letter to the inmate. In addition, staff should maintain a log containing the same information and should request inmates receiving legal mail to sign the log. If the inmate refuses to sign the log, staff should note this refusal in the log.
- Abuse of legal mail privileges—such as using legal mail to transmit contraband, plan escapes, or engage in any of the other activities prohibited under general correspondence provisions (see "General Correspondence," above)—may warrant disciplinary action and the limitation or denial of an attorney's correspondence rights. The warden should refer such matters to the correctional agency's legal counsel.

4. Publications

- Inmates may receive books, magazines, newspapers, and other materials addressed to them (such as advertising brochures and catalogs) without the prior approval of the warden, but these items are subject to inspection and rejection by the institution if they may be detrimental to the security, discipline, or good order of the institution or if they might facilitate criminal activities.
- Because it may be easier to conceal contraband in hardcover publications (such as books) and in newspapers (especially those that are bulky, folded in multiple places, and wrapped), inmates may receive such materials only if they come directly from the publisher, a book club, or a bookstore. Inmates may receive softcover publications (such as magazines, newspaper clippings, and paperback books) from any source. All incoming publications, whether hardcover or softcover, should be inspected for contraband.
- The warden may reject a publication only if it is determined to be detrimental to the security, discipline, or good order of the institution or if it might facilitate criminal activity. Examples of publications that might be rejected would be those that provide instruction in how to carry out criminal activities, or which describe, encourage, or depict the construction or use of weapons, ammunition, bombs, or incendiary devices; methods of escape from correctional facilities; physical layouts of correctional facilities; procedures for brewing alcoholic beverages or manufacturing drugs; activities that might lead to group disruption or the use of

physical violence; or sexually explicit activities (such as sadomasochism, bestiality, or pedophilia) that could pose a threat to institutional security or facilitate criminal activity.
- Wardens should follow the criteria listed above very closely when considering publications for rejection. They should not reject a publication solely because its content is religious, philosophical, political, social, or sexual, or because its content is unpopular or repugnant. In particular, news, informational, scholarly, or literary material on topics involving sex—such as research or opinions on sexual health, reproductive issues, or gay rights organizations—should not be considered sexually explicit publications subject to rejection. Nor should the warden establish a list of excluded publications. Rather, each individual publication or issue of a periodical must be reviewed prior to any rejection; even the rejection of several issues of a periodical is not sufficient grounds to reject an ongoing subscription to that periodical.
- If a publication is rejected, the warden should advise the inmate in writing of the decision and the reasons for it, with clear indications of which specific articles or materials were considered unacceptable. The warden should also advise the inmate that he or she may appeal the rejection decision through the regular inmate grievance procedure and, for the purposes of facilitating any such appeal, the warden should permit the inmate to review the rejected material (unless the material would convey information that would pose a threat to institution security). In addition, the warden should notify the publisher when a publication is rejected and should advise the publisher that an independent review of the rejection may be obtained by writing to the appropriate executive-level official at the correctional agency's headquarters within fifteen days. In order to give either the inmate or the publisher enough time to file an appeal, the warden should retain the rejected publication for at least fifteen days.
- Before ordering any publication, the inmate should be permitted—at his or her discretion—to speak with a designated staff member in order to obtain an indication of the likelihood of individual issues of that publication being rejected. The inmate would be under no obligation to seek such guidance, and any such guidance rendered would not be binding on the institution, but such informal discussions could prevent disappointments and avert bookkeeping problems if and when a publication is determined to be unacceptable.

5. Postage

- Except as noted below, postage charges are the responsibility of the inmate. Postage rate charts should be posted for inmate inspection in housing units and the mailroom, and the commissary should stock postage stamps (in a variety of denominations, including first-class postage and lesser amounts) for inmates to purchase.
- To prevent problems that could occur if inmates possess items of value in significant amounts, the institution may establish limits on the number or total value of postage stamps that an inmate may purchase from the commissary at any one time, the number of times during any one week that an inmate may purchase postage stamps from the commissary, or the number or total value of postage stamps that an inmate may possess at any time.
- The institution should provide postage stamps free of charge to inmates who do not have sufficient funds to purchase postage stamps. To prevent abuses, however, the number of postage stamps provided in this manner may be limited. Reasonable limitations might include providing up to five postage stamps *per week* for an inmate to use for legal mail (letters to attorneys or the courts) and to mail filings under established inmate grievance procedures. In addition, the institution may provide up to five postage stamps *per month* for inmates without funds to mail general correspondence letters. Staff should monitor the commissary accounts of inmates receiving free postage stamps to ensure that they are not abusing the privilege by depleting their accounts immediately before requesting free stamps and then replenishing their accounts immediately thereafter. Inmates who are found guilty under inmate disciplinary procedures of abusing this privilege may be required to reimburse the government for the postage stamps.

Postage stamps may also be provided at government expense in verified emergency situations for inmates who are unable to provide postage and, in limited amounts (up to three stamps per week, unless special reasons exist to provide more) to holdovers and pretrial commitments.

- Inmate packages that need to be forwarded to another institution due to an administrative action (for example, an inmate transfer) will be considered official mail and mailed at government ex-

pense. Such official mail may not be insured, but if the package is lost or damaged the inmate may file a grievance or tort claim against the institution, or a claim against the U.S. Postal Service.
- Inmates may, at their own expense, send letters and packages by registered mail, certified mail, or insured mail and may request return receipts. Inmates, however, may not be provided with express mail or cash-on-delivery services, may not engage in stamp collecting while confined, and may not send mail via private carriers.
- Inmates should sign for all stamps issued to them (whether they have purchased stamps from the commissary or received them at government expense). A log should be maintained for this purpose. The institution business manager is responsible for all stamps sold through the commissary and should conduct quarterly audits of all stamp supplies and issue logs.
- Inmates are prohibited from receiving postage stamps that are not issued by the institution (or, in special cases, issued by other government agencies) or purchased from the commissary. Any uncancelled stamps or postage-prepaid envelopes that may be sent to an inmate from any other source must be returned to the sender, along with a letter of explanation. Relatives, friends, or others wishing to provide an inmate with funds for postage must follow procedures for depositing funds into the inmate's commissary account.
- Inmate organizations are responsible for purchasing their own postage, under guidelines established by the warden.
- Mail received with postage due should not be accepted. If postage-due mail is tendered to the institution mail officer without collection of that postage, the mail should be processed without further collection action.

6. Procedures

- Inmates must be informed of the institution's correspondence regulations promptly after their admission to the institution. In particular, they should be notified that staff members are authorized to open all incoming correspondence and that all outgoing correspondence is placed in the U.S. mail at the request of the inmate, with the inmate assuming full responsibility for the contents of each piece of outgoing correspondence. Inmates should also be informed that outgoing correspondence contain-

ing threats or extortion demands, or that appears to violate laws in other ways, may be referred to the appropriate law enforcement agencies for possible prosecution.
- At least one mail depository that inmates may use for general correspondence should be established within the institution, and at least one mail depository that inmates may use for special and legal mail should be established.
- Inmates on holdover status (i.e., en route to another institution) should have correspondence privileges similar to those of other inmates, insofar as it is practical to do so. Inmates in segregation should have full correspondence privileges, unless placed on restricted general correspondence status as a classification or disciplinary matter.
- Writing paper and envelopes preprinted with the institution's return address, and containing a space where the inmate is required to place his or her name and registration number, may be provided at no cost to the inmate. Inmates who use their own envelopes must place a return address on the envelope, containing their name and registration number, P.O. box, city, state, and ZIP code.
- Ordinarily, when incoming correspondence is rejected, the warden should notify both the sender and the inmate, advise them of the reasons for the rejection, and notify them of their rights to appeal the rejection; the warden should also return the letter to the sender. However, if the correspondence is rejected because it contains evidence of a crime or includes plans for or discussion of the commission of a crime, then it is unnecessary to provide notification of the rejection. Further, such correspondence should be referred to appropriate law enforcement agencies instead of being returned to the sender.
- Friends or family of an inmate may mail funds (in the form of checks or money orders) to be credited to the inmate's commissary account, provided the inmate completes appropriate forms and the negotiable instruments are properly prepared. Negotiable instruments that are not properly prepared should be returned to the sender with a letter of explanation, and a copy of that letter should be sent to the inmate.
- If outgoing mail sent by inmates is undeliverable, it will be returned to the institution by the U.S. Postal Service. In such cases, staff members must verify that the returned correspondence did indeed originate with the inmate, must ensure that the returned correspondence was not tampered with before being

returned, and must check for contraband. Therefore, general correspondence that is returned to the institution should be opened and inspected. Legal mail and special mail must be opened and inspected in the presence of the inmate.
- Appropriate U.S. Postal Service change-of-address forms and cards should be made available to inmates who are being transferred to other institutions or released. Inmates are responsible for notifying correspondents of their address changes and for postage for mailing change-of-address cards.
- The staff should forward incoming special mail and legal mail to transferred or released inmates whenever practicable. The staff should forward incoming general correspondence for a thirty-day period after an inmate's transfer or release. Correspondence that cannot be forwarded should be returned to the sender. Inmates who have been released on a writ or for other purposes for thirty days or less should be given the option of having their correspondence forwarded to them or kept at the institution pending their return; if such an inmate refuses to elect one of these options, that fact should be noted in his or her central file and all general correspondence for that inmate should be returned to the U.S. Postal Service as undeliverable.

III. VISITING REGULATIONS

Overview

Visiting by family, friends, and members of community groups can help maintain inmate morale and strengthen relationships between inmates and their family members or others in the community. Therefore, visiting is a sound correctional management feature that should be encouraged; nevertheless, visiting may be subject to certain restrictions, in order to maintain institutional security.

1. Visitors Lists

- During the admission and orientation process, staff members should ask each inmate to submit a list of proposed visitors. Following appropriate investigation of the proposed visitors, an approved visitors list should be provided to the inmate and to the visiting room officer.

The list may be amended at any time, in accordance with the procedures cited below. The inmate's approved visitors list must be updated immediately, whenever names are added or deleted, with copies of the revised list being distributed to the inmate and to the visiting room officer. Visitors lists should also be maintained in the inmate's central file.

- Staff members may request background information from potential visitors who are not members of the inmate's immediate family before placing them on the approved visitors list. When little or no information is available on a potential visitor, visiting privileges may be denied. Because of the heightened security needs that exist in medium- or high-security institutions, background information should be obtained on all potential visitors who are not members of the immediate family. Exceptions to these information-gathering procedures may be made for pretrial inmates and under other appropriate circumstances.
- Proposed visitors may be requested to complete a visitor inquiry form. The signature of a parent or guardian on this form is necessary to process a request for a proposed visitor under eighteen years of age.
- If a more detailed background investigation is necessary before approving a visitor, the inmate may be responsible for forwarding a release authorization form to the proposed visitor. That form must be signed and returned to the staff by the proposed visitor before further action will be taken. Upon receipt of the authorization form, the staff may forward a questionnaire (along with a copy of the authorization form) to appropriate law enforcement or crime information agencies for a criminal background check. Documentation on visitors should be retained in the inmate's central file.
- The staff should notify the inmate of each approval or disapproval of a proposed visitor. Upon approval of a visitor, the staff should provide the inmate with a copy of the visiting guidelines and directions for transportation to and from the institution. The visiting guidelines should cite specific statutes prohibiting the introduction of contraband into the institution, and cite the criminal penalties for attempting to do so. The inmate is responsible for notifying the visitor of the approval or disapproval and is expected to send the visiting guidelines and transportation directions to the approved visitor.

2. Categories of Visitors

- Regular visitors are those on the approved visitors list. The list may include members of the inmate's immediate family (parents, stepparents, foster parents, siblings, spouse or common law spouse, children, and stepchildren). Immediate family members ordinarily should be approved for inclusion on the visitors list, unless there are unusually strong reasons for disapproving them. Other relatives, including grandparents, uncles, aunts, nieces, nephews, and in-laws may be placed on the approved list if no reasons exist to exclude them.
- Other regular visitors on the approved visitors list could include friends and associates of the inmate. In general, an inmate's approved visitors list should not include more than ten friends and associates, although the warden may grant exceptions where warranted.

At minimum- and low-security institutions, visiting privileges ordinarily should be extended to friends and other nonrelatives, unless their visits could reasonably be expected to create a threat to the security and good order of the institution. At medium- and high-security institutions, visiting privileges should be extended only to friends and associates having an established relationship with the inmate prior to confinement; as with friends and associates of inmates at lower-security facilities, visiting privileges should only be extended if there is no threat to the security and good order of the institution. The warden may make exceptions to the prior relationship rule, particularly for inmates without other visitors, if it can be shown that the proposed visitor is reliable and poses no threat to the institutional security.

Regular visitors lists may also include representatives from community groups, such as civic and religious organizations, for the purpose of providing approved services. This category, however, would not include past or present participants in an institution's volunteer and community involvement program, who ordinarily should not be approved for inclusion on an inmate's visitors list.

- Apart from regular visitors, the warden may authorize special visits from various categories of visitors. Although inmates should not be permitted to engage actively in a business or profession, and should have assigned authority for the operation of any business or professional enterprises to others for the duration of their confinement, there may be occasions where a decision must be made that will substantially affect the assets or prospects of an inmate's business. In such cases, the warden may permit a special visit from a business associate.

If an inmate is a citizen of a foreign country, then the warden must permit consular representatives of that country to visit the inmate on matters of legitimate business. This privilege may not be withheld, even if the inmate is in disciplinary status.

Clergy, former or prospective employers, sponsors, and parole advisors may be approved as special visitors, for purposes such as assisting in release planning, counseling, and discussion of family problems.

Attorneys should be approved as special visitors and, whenever possible, attorney visits should take place in private conference rooms or in regular visiting rooms where the inmate and his or her attorney may be afforded a degree of privacy and separation from other inmates. Staff members must never be permitted to subject visits between an attorney and an inmate to auditory supervision.

Finally, the warden may approve regular visitors as special visitors if they need to visit during other than regularly established visiting times.

- In rare instances, sentences or court orders may provide that an inmate not associate with a specified person, owing to probable cause that such an association would be for the purpose of enabling the inmate to participate in an illegal enterprise. The correctional agency's legal counsel should advise the staff at individual institutions on how such sentences or court orders would affect an inmate's visiting privileges.
- If a proposed visitor has a criminal conviction on his or her record, the staff should consider the nature and extent of any convictions, how recently they occurred, and how the presence of someone with such convictions might affect institution security before rendering a decision on whether that person should be approved for visiting. Specific approval of the warden may be necessary before a person with a criminal record might be permitted into the institution as a visitor. Written authorization from probation and parole officials should be obtained before an individual on probation, parole, or supervised release may be added to an inmate's approved visitors list.

- Unless an exception is made by the warden, children under the age of sixteen may not visit unless accompanied by a responsible adult. Children should be kept under the supervision of that adult visitor or a children's program at the institution. Children who are not properly supervised may be excluded from visiting.

3. Visiting Rooms, Schedules, and Procedures

- One or more visiting rooms at each institution should be arranged so as to provide adequate supervision, adapted to the degree of security required by the type of institution. The visiting area should be as comfortable and pleasant as practicable, and appropriately furnished.

If space is available, a portion of the visiting room may be set up to provide facilities for visitors' children and another portion of the visiting room may have vending machines where inmates and their visitors may purchase soft drinks, snacks, sandwiches, etc. Restrooms and pay telephone service should be available for visitors. Directions for transportation to and from the institution and the telephone numbers of commercial transportation companies (taxicabs, buses, etc.) should be posted for visitors to see.

Minimum- and low-security institutions may permit visits beyond the secure perimeter, but always under the close supervision of staff. These institutions may establish outdoor visiting areas, but they must always be inside the secure perimeter and under staff supervision.

- At a minimum, visiting hours should be established on Saturdays, Sundays, and holidays. Because restricting visiting to those days may cause hardships for some families, other suitable hours, including evening visiting hours, may be established if resources permit.
- Each inmate should be allowed a minimum of four hours of visiting time per month, but the length and frequency of visits may be limited to avoid chronic overcrowding of the visiting area and to meet security needs and available resources. For example, an inmate may be permitted to receive visitors on only one of the weekend visiting days instead of both, guidelines may be established setting a maximum number of visitors an inmate may see at one time, and the length of individual visits may be limited.

Exceptions may be made to any such guidelines, however, given special circumstances such as the distance a visitor must travel, the inability of a visitor to visit on a particular day or time, the frequency with which an inmate receives visitors, the size of an inmate's family, or health problems that an inmate or visitor may have.

- Staff should verify the identity of each visitor against a driver's license or other form of photographic identification card prior to admitting the visitor into the institution. The usual means of identification, however, need not be the sole basis of identification, and tactful questioning on the basis of available information may resolve doubtful cases.
- The staff should make available to all visitors written guidelines for visiting the institution and should request the visitor sign a statement acknowledging that the guidelines were provided and that the visitor does not have any article in his or her possession that would threaten the security of the institution. Visitors who refuse to sign such a declaration may be denied visiting privileges. The staff may require a visitor to submit to a personal search, including a search of any items of personal property, as a condition of allowing or continuing a visit.
- A log of all visitors to the institution should be maintained, and each visitor must sign either the log or another official record retained by the institution to document the day and time of his or her visit.
- Inmate visits should be supervised by staff in order to prevent the passage of contraband and to ensure institutional security and good order. Even visitor restrooms may be monitored (by a staff member of the same gender as the visitor using the restroom), but only with the warden's approval and only if there is a reasonable suspicion that a visitor or inmate may engage or be engaging in some form of prohibited behavior. Notices should be posted informing visitors of the potential for monitoring anywhere in the visiting area.
- The visiting room officer is responsible for ensuring that all visits are conducted in a quiet, orderly, and dignified manner. This officer, in consultation with the shift commander or duty officer, may terminate visits that are not conducted in an appropriate manner.
- At some institutions "contact visiting" is permitted, whereby inmates and their visitors may shake hands, embrace, or kiss, but

only at the beginning and the end of each visit and only within the bounds of good taste. The staff may limit physical contact to minimize opportunities for the introduction of contraband and to maintain the orderly operation of the visiting area. Inmates in Special Management Units may be allowed "noncontact" visiting privileges only.
- Staff may not accept articles or gifts of any kind for an inmate, except packages that have been approved by the warden or designated staff member. The warden may permit a visitor to leave money with a designated staff member for deposit in the inmate's commissary account, provided that proper documentation is kept to prove that the money was properly tendered and properly deposited.
- The visiting room officer must be aware of any articles passed between the visitor and the inmate. If there is any reasonable basis to suspect that an item may be contraband, the visiting room officer may examine it.
- Violations of visiting guidelines may result in disciplinary action against the inmate, which may result in the denial of future visits. Moreover, criminal prosecution may be initiated against the visitor, the inmate, or both in the case of criminal violations.
- Visitors are not permitted to bring animals onto institution grounds, except for dogs that assist persons with disabilities.

4. Visits to Inmates in Special Housing Status

- Visiting privileges for inmates in holdover status or undergoing admission and orientation may be limited to immediate family members, especially in cases where there is neither a visiting list from a transferring institution nor other verification of proposed visitors.
- Inmates hospitalized at the institution may receive visitors at the discretion of the chief medical officer, in consultation with the chief of security. If the chief medical officer recommends against a visit because the inmate is suffering from an infectious disease, is in a psychotic or emotional episode that makes visiting inadvisable, or is otherwise not in a condition to receive visitors, then the situation should be explained to the proposed visitor in a careful and sensitive manner and documented in the inmate's central file. Visits to inmates hospitalized in the outside community may be restricted to immediate family, are subject to the general

visiting policies of the hospital, and must be approved by the chief of security.
- An inmate in administrative segregation or disciplinary detention may receive visits in accordance with the same policies and regulations that apply to general population inmates, providing such visits do not pose a threat to the security or orderly operation of the institution, unless the inmate has been denied visiting privileges as a disciplinary measure, in accordance with established disciplinary procedures. For inmates in administrative segregation or disciplinary detention who have not been denied visiting privileges, the warden may authorize special visiting procedures to avert any threats to security or good order.

IV. INMATE GRIEVANCE PROCEDURES

Overview

Having an established inmate grievance procedure in place can enable inmates to obtain formal reviews of complaints relating to any aspect of their imprisonment. Not only can such a procedure bring about a redress of legitimate grievances, it can also avert unnecessary litigation and serve to maintain institution morale by providing a tangible method for alleviating pressure and responding to inmate concerns.

The grievance review process should be open to any inmate in an institution operated by the correctional agency or in any halfway house or privately managed prison under contract to the agency or otherwise managed under the agency's responsibility. Inmates may be permitted to file grievances relating specifically to their conditions of confinement or other aspects of their imprisonment that would be the responsibility of the correctional agency. Filings should not be accepted, however, for tort claims or accident compensation claims, which would be handled by the courts or other agencies. Nor should grievance filings be accepted on matters relating to criminal convictions or sentencing procedures (except for sentence computations performed by staff members of the correctional agency) or relating to complaints made on behalf of other inmates. Inmates confined under contract in prisons operated by the correctional agencies of other states ordinarily should be subject to grievance procedures available in those other agencies.

1. Responsibility

- At institutions, initial filings of formal grievances are reviewed by the warden; at halfway houses, initial filings of formal grievances are reviewed by the community programs manager. Appeals of rulings made by the warden or the community programs manager are reviewed by an appropriate executive-level official at the correctional agency's headquarters—usually the agency's legal counsel or the assistant director or assistant commissioner responsible for correctional programs.
- Responsibility at the institution level for coordinating grievance procedure operations may be delegated to a staff member above the department head level. Responsibility at the headquarters level for reviewing appeals may be delegated to an inmate appeals administrator.
- At each stage in the grievance process, the responsible official must ensure that formal inmate complaints or appeals are (a) acknowledged with a written receipt; (b) investigated; and (c) given a formal response, following the investigation, review, and ruling.

2. Forms

- The correctional agency should develop two types of forms to be used in the grievance process: one for filing initial grievances and one for filing appeals. Both would include a space where the inmate may write the specific complaint, a tear-off sheet for acknowledging receipt of the complaint or appeal, and a space where staff members may enter the official findings and ruling on the complaint or appeal. In addition, there should be a space on all parts of the form for the name, registration number, and institution of the inmate, and for a case number for the complaint (the case number for the appeal would be the same as the case number for an initial complaint). Both forms should have multiple copies (using carbon paper or other method), so that copies may go to the inmate, to a centralized grievance file at the institution or halfway house, and, as appropriate, to a centralized appeals file at headquarters. Copies of grievance paperwork should not be placed in the inmate's central file. Grievances and appeals should be filed by case number at both the institution level and at headquarters.

The forms should be provided upon request to inmates by correctional counselors.

3. Procedures

First Step: Informal Resolution

Before filing a formal complaint, an inmate should present any complaints informally to the staff and the staff should attempt to resolve the complaint informally. The warden may establish local procedures to ensure that attempts at informal resolution are made. Inmates in halfway houses would not be required to attempt informal resolution before filing a complaint.

Second Step: Filing the Complaint

If a complaint cannot be resolved informally, the inmate would have fifteen days to obtain a grievance form from his or her counselor (or appropriate staff at a halfway house). That form should be used for submitting a formal complaint to the warden or community programs manager (as appropriate). Inmates may be requested (although not required) to submit separate grievance forms for each specific complaint. Complaints must be filed using the appropriate form or they will not be accepted. The warden or community programs manager may grant extensions to the fifteen-day filing period if there are valid reasons for delay.

After completing the grievance form, the inmate should submit it to his or her counselor, who will immediately forward it to the warden or to the staff member delegated to coordinate inmate grievance procedure operations at the institution.

An inmate may receive assistance from other inmates or from institution staff in preparing a complaint. In particular, the institution should have procedures in place for assisting incapacitated, illiterate, or non–English-speaking inmates to prepare complaints.

If an inmate believes that a complaint is sensitive, and that he or she would be adversely affected if the complaint became known at the institution, then it is permissible to file the complaint directly with the correctional agency's headquarters. Such complaints should include the inmate's written explanation for not filing the complaint at the institution or halfway house level. If the responsible headquarters staff agrees that the complaint is sensitive, then it should be accepted and reviewed at that level; if the headquarters staff believes that the complaint is not sensitive, then the inmate should be advised that the complaint will not be reviewed at headquarters. If the headquarters staff does not accept an initial complaint for review, then the inmate has the option of filing the complaint with the warden or community programs manager, as appropriate.

Initial grievances concerning placement in a Special Management Unit may be filed directly with headquarters.

Responses to complaints filed at the institution or halfway house level must be made within fifteen days of the filing; responses to complaints filed at the headquarters level must be made within thirty days. Extensions of up to fifteen days may be granted for good cause for complaints at the institution or halfway house level or thirty days at the headquarters level. If the complaint is of an emergency nature, and the inmate's immediate health or welfare may be in jeopardy, the response should be made within forty-eight hours of the filing.

Staff who participated in any stage of processing a disciplinary action against an inmate may not be involved in investigating an appeal of that action, and complaints about staff misconduct must not be investigated by any staff member who is the subject of such complaints.

The official response (typewritten in the appropriate section of the grievance form) should summarize the initial complaint, answer the complaint, cite appropriate agency policies or regulations, and be written in such a way that it can be released to any inmate and to the general public under applicable Freedom of Information or privacy statutes.

Third Step: Appeal

If an inmate is not satisfied with the response of the warden or the community programs manager, he or she has thirty days to file an appeal with headquarters. The appeal should be written on the appropriate form (obtained from the inmate's correctional counselor) and, once filed, should be answered by headquarters within thirty days. The general procedures and requirements for filing, investigating, and responding to an appeal would be comparable to procedures and requirements for the initial complaint.

The ruling from headquarters on the appeal would represent the final administrative response to a complaint from the correctional agency. Inmates who wish to carry their complaints further may do so only through the courts.

V. INMATE LEGAL ACTIVITIES

Overview

Each institution should afford inmates reasonable access to legal materials and counsel and reasonable opportunity to prepare legal documents. At least one inmate law library should be established at each institution and procedures should be in place to enable inmates to confer with their attorneys in person, by telephone, or though correspondence. Inmates should be permitted to have legal materials in their cells and to have time to conduct legal research and prepare legal documents.

1. Law Libraries

- The correctional agency's legal counsel should prepare a list of required law library materials and update that list in June and December of each year with an inventory checklist of required law library materials. All the materials specified should be maintained in each institution's main inmate law library. Volumes that are misplaced or destroyed must be replaced within a reasonable time after the staff become aware of the loss.
- In addition to main law libraries, some institutions may establish smaller (or basic) law libraries at satellite camps or in certain housing units—such as Special Management Units—for inmates who may have difficulty visiting the main law library. In addition, basic law libraries may operate in place of main libraries at institutions whose inmate populations are below a specified level. Basic law libraries would not hold as many volumes as main law libraries and would draw on a separate list of required materials for basic law libraries prepared by the correctional agency's legal counsel. Basic law libraries may borrow volumes not in their collection from the institution's main law library as needed.
- The legal counsel's office should provide directions on where to order specific publications (names and addresses of publishers, etc.), and the agency's budget administration department should ensure that funds are available for the purchase of inmate legal materials and should establish procedures whereby institutions can procure such materials.
- The supervisor of education (or whichever official supervises library activities at the institution) should be responsible for the main inmate law library and for any basic law libraries that might

also be established. He or she should ensure that all materials specified on the list prepared by the agency's legal counsel are in the libraries' collections. Upon receiving the inventory checklist from the agency's legal counsel every June and December, the supervisor of education should have thirty days to conduct an inventory against the checklist to identify missing or damaged materials. Thereafter, the supervisor of education should have thirty days to order replacement materials for those that are missing or damaged and to order copies of other materials that have been added to the required list since the previous inventory. A copy of the inventory checklist, noting all missing or damaged materials, should be sent to the agency's legal counsel.

- In addition to obtaining inmate legal materials through government purchases, law libraries at institutions may acquire materials through donation (with the specific approval of the agency's legal counsel). In addition, if inmates from another jurisdiction are being held at an institution on a contact basis, the correctional agency or other governmental body in that jurisdiction should provide legal materials for use by those inmates. For example, if a state prison houses federal offenders on a contract basis, then the federal government should provide necessary publications on federal laws, regulations, and court cases for those inmates to consult.
- The main law library should be situated in a large room where law books and publications may be shelved and with space for tables where inmates may work with the legal materials. There should be sufficient space for expansion of the collection, which will occur as each new legal reporter volume is received and as other materials are added to the list of required publications issued by the agency's legal counsel.
- The education supervisor should be responsible for supervising the main law library and any basic law libraries. Given the expense of replacing law library materials, the staff must monitor activities in law libraries carefully in order to prevent damage or theft of materials. Supplements to the loose-leaf *Criminal Law Reporter* should be interfiled as soon as they are received, and out-of-date supplements and other outdated materials should be disposed of as soon as replacement materials are received.
- Inmates may be given work assignments to catalog, update, and shelve legal materials but should not be responsible for conducting inventories, ordering publications, or otherwise supervising institution law libraries.

- Institution copying equipment or copy vending machines may be made available for inmates to reproduce materials needed for research outside the library. Procedures should be established whereby an inmate may request a reasonable amount of reproduced material and indicate why being able to review original material in the library is not adequate for his or her needs. By providing ample hours for library usage and by making copying equipment available, the institution should be able to reduce the incidence of mutilation or theft of library materials.
- Unauthorized possession of law library materials by an inmate constitutes a prohibited act and may warrant disciplinary action.

2. Legal Research and Preparation of Legal Documents

- Inmates may be permitted ample time to conduct legal research and prepare legal documents. Where practical, materials in the inmate law library should be available during evening and weekend hours, and inmates may be allowed to prepare legal documents in their living quarters. Ordinarily, inmates should carry out such activities in their leisure time, but special time allowances may be granted under certain circumstances. For example, if an inmate is facing an imminent court deadline and his or her available leisure time is not sufficient to conduct research and prepare necessary documents, then the warden may authorize the inmate to work fewer hours on his or her job assignment. Except in such unusual situations, however, inmates should conduct legal research and work on legal documents in such a way that does not interfere with their regular institutional activities.
- Inmates may be permitted to retain legal materials in their living quarters, in reasonable amounts. Local regulations should specify the quantity permitted, and limitations should be governed mainly by housekeeping considerations. Inmates should not be permitted to retain personal legal materials in such quantities as to pose fire, sanitation, or security hazards.

Inmates may obtain books and other legal materials in accordance with established policy for receiving publications (see Part II, "Publications," above) and retaining them in their living quarters.

Inmates may receive court papers or other legal documents from court clerks, judges, or their attorneys that are mailed or given to them by visitors. Staff members may inspect these documents for contraband but may not read them if they are properly presented. The warden must ensure that special care is taken in rejecting or limiting any documents provided by courts or attorneys, and the staff must be aware that they are never authorized to make a determination that the content of specific material sought by an inmate is or is not relevant to the inmate's case. Instead, materials may be disallowed only if there is a clear and compelling reason to believe that they would cause a security or other hazard, and only upon consultation with the correctional agency's legal counsel.

Inmates confined in disciplinary detention or administrative segregation may retain up to one cubic foot of legal materials in most cases (with excess amounts being stored, along with excess personal property, for the duration of their confinement in that status). Greater amounts may be permitted, however, if an inmate has an imminent court date and can demonstrate a need for more materials.

- Unless it is clearly impractical to do so, inmates may be permitted to use typewriters to prepare legal materials. If an inmate cannot type, he or she may have another inmate type legal documents. Inmates, at their own expense, may also be permitted to hire stenographers to type documents outside the institution.
- Unless the institution has an active, ongoing legal aid program for inmates (see "Legal Aid Programs," below), the warden should allow an inmate to seek the assistance of another inmate, during leisure hours, for purposes of conducting legal research and preparing legal documents. Limitations may be placed on such assistance in the interest of institution security, good order, or discipline.
- Each inmate is responsible for submitting his or her own legal documents to court. Institution staff who are authorized to administer oaths should be available to provide necessary witnessing of those documents, as requested by inmates and at times scheduled by staff.
- Ordinarily, inmates should use carbon paper to make copies of the legal documents they prepare. If they can demonstrate that they need several copies of legal documents for submission to the courts, they may be permitted to make copies using the copy vending machine at the law library (if one has been placed there), or they may, at their own expense, request the staff to copy such documents on an institution copy machine. Such services will be provided, but only if they do not interfere with regular institution operations.

The cost of making copies may be waived if the inmate is without funds or if the amount of copying is minimal. Staff members should monitor the commissary accounts of inmates requesting copying at government expense due to lack of funds to ensure that they are not abusing this privilege. If it appears that an inmate is following a pattern of depleting his or her commissary account before requesting copies of legal documents at government expense, and then replenishes the account immediately thereafter, that inmate may be required to reimburse the government for any copying.

3. Retention of Legal Counsel by Inmates

- Inmates have the right to contact and retain attorneys. With the written consent of the inmate, staff members may advise an attorney of the inmate's available funds. The staff may not interfere with an inmate's selection and retention of attorneys if that inmate has attained majority and is mentally competent, and if the attorney selected is a member of the bar. If the inmate is mentally incompetent or a minor, the warden should refer all matters concerning the retention and payment of attorneys to the inmate's guardian or to the courts. The correctional agency cannot act as a guarantor or collector of attorney's fees.
- Visits by retained, appointed, or prospective attorneys of an inmate must be permitted; visits by an attorney who wishes to interview another inmate as a witness must also be permitted, if that inmate is willing to meet with the attorney.
- The frequency of attorney visits may not be limited, because the number of visits is dependent upon the nature and urgency of the legal problems involved. The warden, however, may set the time and place for visits and require that they ordinarily occur during regular visiting hours. Exceptions may be granted according to local conditions or if the inmate or attorney can demonstrate that an emergency situation exists.
- Attorneys should contact the warden to make appointments for visits with inmates prior to each visit. When prior notification is not practical, however, the warden should make every effort to accommodate the visit.
- Prior to visiting an inmate, an attorney should be required to present identification and confirm that the purpose of the visit is to confer with an inmate that he or she represents, to confer with an inmate who has requested a visit, or to interview an inmate as a witness. The attorney must not be required to state the subject

matter of the litigation or the interview. In addition, an attorney may be asked to provide verification that he or she is licensed to practice law and is an attorney in good standing. If questions arise concerning an individual's qualifications in this regard, the institution should refer the matter to the correctional agency's legal counsel. Attorneys may be subject at any time to a search of his or her person and belongings for the purpose of detecting contraband.
- Attorney visits should take place in a private conference room, if one is available, or in a regular visiting room in an area or at a time conducive to privacy. Staff are forbidden to subject visits between an inmate and an attorney to auditory supervision.
- Attorneys may be permitted to make tape recordings of visits, if they state in writing in advance of the interview that the sole purpose of the recording is to facilitate the attorney-client or attorney-witness relationship. The use of other electronic devices, such as videotape recorders or computers, ordinarily should not be permitted. The warden may make exceptions if it can be shown that the use of such equipment is essential to facilitate the attorney-client relationship and that it would not jeopardize the security, discipline, or good order of the institution.
- Limitation or denial of an attorney's privileged visitation and correspondence rights may be imposed if an attorney misrepresents his or her identity or qualifications; introduces (or plans to introduce) contraband into the institution; conspires, commits, or attempts to commit an act of violence or disruption within the institution; encourages an inmate to violate the law or correctional agency regulations; or engages in any other activities that violate regulations or threaten the security, good order, or discipline of the institution. Unless the breach of regulations is extreme, limitation rather than full denial of visitation or correspondence rights is proper, especially in cases where the inmate is represented by the attorney in question and is confronted with a court deadline. It is advisable for the warden to consult with the correctional agency's legal counsel before imposing any limitations or denials.
- An attorney may appeal any limitation or denial of visiting or correspondence rights to the commissioner or director of the correctional agency. The inmate may appeal by filing a complaint under established inmate grievance procedures.
- Paralegals, clerks, and legal assistants may be accorded the same status as attorneys with respect to visiting and correspondence,

provided they are working under the supervision of an attorney. The attorney must certify the assistant's ability to carry out his or her tasks, pledge to supervise the assistant's activities, and accept personal and professional responsibility for all acts of the assistant that may affect the institution, the inmates, and staff. In addition, the assistant may be required to complete and sign a personal history statement and agree to abide by agency regulations and institution guidelines. The warden, however, may prohibit a paralegal, clerk, or legal assistant from visiting or corresponding with an inmate if such action is necessary to maintain the security, discipline, or good order of the institution.

4. Legal Aid Programs

- If the correctional agency approves and funds a legal aid program that provides a broad range of legal assistance to inmates, then the staff should allow that program to operate with the same independence as privately retained attorneys.
- Law students or legal assistants working in the legal aid program must be supervised by attorneys or law school professors, under the same terms as paralegals, clerks, and legal assistants working for privately retained attorneys. They should be granted the same status as attorneys, with respect to visiting and correspondence rights.
- If the warden believes it is appropriate to terminate or restrict a program or to deny visitation or correspondence privileges to individual participants, then he or she must refer the matter to the correctional agency's legal counsel.

VI. NEWS MEDIA CONTACTS

Overview

The correctional agency is obligated to keep the public informed about matters relating to its institutions and operations, and it should recognize the right of the news media to seek access to inmates for interviews and the right of inmates to speak with members of the news media. At the same time, the correctional agency is obligated to protect the privacy of inmates. An institution's news media policy seeks to balance these competing interests, while also preserving the security, discipline, and good order of the institution.

This section pertains exclusively to news media policy that relates directly to inmate access to the news media and to inmate privacy concerns; it does not pertain to policies on news media interviews with staff, the development of media strategies, or the appointment of official spokespersons for the correctional agency or its individual institutions.

1. Inmate Interviews and Institution Visits

- A news media representative who desires to visit the institution or conduct an interview at the institution must apply in writing to the warden, indicating that he or she is familiar with the policies and guidelines of the institution and agrees to comply with them. Failure to adhere to the standards of conduct set forth in institution policies and guidelines would constitute grounds for denying that news media representative, or the news organization that he or she represents, permission to conduct an interview. The correctional agency may wish to place conditions on interviews, such as requiring the news media representative to give the agency an opportunity to respond to allegations or prohibiting the news media representative from gathering personal information from one inmate about another inmate who declines to be interviewed.
- Actual appointments for visits and interviews must be made in advance. Either an inmate or a representative of the news media may initiate a request for a personal interview at the institution. Inmates may not receive compensation or anything of value for interviews with the news media, may not be employed to act as a reporter, and may not publish under a byline.
- Staff should notify an inmate of any interview request that is submitted by a representative of the news media. The inmate must give written consent for the interview before the interview can take place. A copy of the inmate's written consent or denial should be placed in the inmate's central file. As a prerequisite to granting an interview, the inmate must authorize the staff to respond to comments made in the interview and to release information to the news media concerning any comments made by the inmate in the interview.
- Once an inmate has agreed to an interview, the warden should decide—normally within forty-eight hours—whether to permit

the interview. The warden may deny the interview if the news media representative, or the news organization that he or she represents, does not agree to abide by institution guidelines or has failed to abide by them in the past; if the inmate is physically or mentally unable to participate; if the inmate is a juvenile and written consent has not been obtained from his or her parent or guardian; if the interview, in the opinion of the warden, could endanger the health or safety of the interviewer or disturb the good order of the institution; if the inmate is involved in a pending court action and the court has issued an order forbidding media interviews; if the inmate is not convicted (e.g., a pretrial detainee) and requests for interviews must be cleared with the court or the prosecutor's office; or if the inmate is a protection or special witness case whose safety would be endangered if his or her whereabouts were revealed.

- Interviews should be held in the institution visiting room during regular weekday business hours and normally should not be subject to auditory supervision by staff members. The warden has the discretion to determine that another location would be more suitable for an interview, to limit interview times if interviews are causing a serious drain on staff or facilities, to limit interviews to one hour per month for inmates in special housing or hospital status, and to limit the amount of audio, video, and film equipment and the number of media personnel entering the institution. In addition, the warden may suspend all media visits during an institutional emergency and for a reasonable time after an emergency.
- If a media representative wishes to tour the institution (whether or not the tour would be in conjunction with an inmate interview), specific permission must be requested. When media representatives visit institutions, they may meet with groups of inmates engaged in authorized programs and activities and may take photographs of those programs and activities. Inmates have the right not to be photographed and not to have their voices recorded by the media. Therefore, a visiting media representative is required to obtain written permission from an inmate before photographing that inmate or recording that inmate's voice. A copy of that written authorization should be placed in the inmate's central file.
- The warden may establish a press pool whenever he or she determines that the frequency of requests for interviews and visits reaches a volume that warrants limitations. Members of the press

pool are selected by their peers and consist of not more than one representative from each of the following groups: the national and international news services, the television and radio networks and outlets, news magazines and newspapers, and local media in the community where the institution is located. All news material collected by such a press pool should be made available to all media, without the right of first publication or broadcast.

2. Release of Information

- Unless authorized by the inmate to release additional information, the correctional agency may provide the news media only with information about an inmate that is a matter of public record, including the inmate's name, registration number, age, race, conviction and sentencing data, past movement via transfers or writs, general institutional assignments, and place of incarceration (the place of incarceration may be released only after the inmate has arrived there; the release of designation information is prohibited).
- The information cited above may not be released if the inmate is a protection or special witness case and the information is considered confidential.

CHAPTER 4

Programs and Services

I. INMATE EMPLOYMENT, EDUCATIONAL PROGRAMS, AND VOCATIONAL TRAINING

Overview

Work assignments, educational programs, and vocational training for inmates can provide several benefits. Most notably, they can offer inmates opportunities for constructive and socially useful self-improvement that can enhance their prospects for securing honest employment and succeeding as law-abiding citizens after they are released and return to the community. Further, such programs alleviate inmate idleness, which is one of the most serious challenges to security and good order in an institutional environment.

1. Inmate Employment

- If authorized by state or federal law (depending on the jurisdiction), inmates may be employed on various work projects within an institution and receive nominal pay for the work they perform. Pay rates may be set or authorized by enabling legislation or by correctional agency regulations and should adhere to a pay table or schedule that sets forth higher or lower hourly wages, depending upon the type of work being performed and length of successful service in a particular position. Inmates may also earn and accumulate paid leave or vacation time (such leave or vacation time, of course, would not affect an inmate's period of incarcera-

tion but would enable the inmate to take time off from work to engage in leisure activities within the institution).
- Wages earned through prison work assignments ordinarily should be credited directly to the inmate's commissary account. In addition, statutes in a given jurisdiction may provide for the deduction of a portion of an inmate's earnings in order to defray costs of confinement or to pay court-ordered fines, child support, victim restitution, or other obligations.
- Work assignments may be required for all inmates who are physically able to work. Work assignments may be changed, and inmates may be given less desirable assignments as a disciplinary measure.
- Although work assignments are intended to reduce inmate idleness and provide services for institutional support or other purposes, they also may be seen as providing inmates with valuable work experiences, job skills, and on-the-job training.
- Work assignments may be institutional, public service, or industrial in nature. Institutional assignments support institutional operations and would include janitorial services, clerical services, farming (to produce food for the institution mess halls), food preparation, carpentry, painting, plumbing, groundskeeping, and routine maintenance and repair work. Public service assignments generally involve work on community or government-operated projects outside the institution, such as road building, groundskeeping and general maintenance at public parks and other government-owned lands, and groundskeeping on military bases. Industrial operations include the production of goods, generally in a factory or workshop within the institution, such as road signs, license plates, office furniture, military uniforms, military blankets, and printed materials.

In most jurisdictions, industrial work assignments are most likely to be subject to legislation regarding goals, products, and operations. It is not unusual for legislation to require that prison industrial operations be set up in such a way as to avoid undue competition with the private sector. Prison industrial operations, typically, must be diversified (in order to minimize their market share in any one industry), labor-intensive (to ensure that they are achieving their main purpose of reducing inmate idleness and to prevent utilizing technology in such a way as to increase competitiveness against the private sector), and subject to guidance from boards of directors representing all sectors of the economy. Typically, the principal statutory restriction on prison industrial enterprises is that they produce goods solely for sale to and use by government agencies. It is common for laws to prohibit the sale of prison-made goods on the open market. Correctional agencies and institutions should work closely with their legal counsel to ensure that any industrial programs they establish conform to both the letter and the spirit of the enabling legislation, and that they avoid putting undue pressure on private sector business enterprises.

- Inmate work sites should comply with applicable regulations of the federal Occupational Safety and Health Administration (OSHA).
- Administrators of inmate work programs should confer with security administrators at all times to ensure that inmates are given appropriate supervision while performing work assignments and that all necessary security protocols are in place and are being observed. For example, outside work details—such as maintenance crews doing groundskeeping outside the secure perimeter, or road construction crews working in the community—generally should be limited to low- or minimum-security inmates. As another example, a variety of security methods should be employed in factories, in order to prevent tools or machine parts from being stolen for use as weapons. All areas of factories, including areas where raw materials or finished products are stored in bulk, must be searched regularly for contraband and weapons; tool control should be practiced intensively, with shadow boards for the placement of all tools and the use of metal detectors to ensure that tools are properly accounted for and to reduce opportunities for inmates to steal them.

2. Types of Education and Training Programs

- The warden should ensure that each inmate with the need and capacity to do so, and sufficient time remaining on his or her sentence, has the opportunity to participate in one or more of the educational and vocational training courses.

Adult Literacy Programs

Those inmates who have not already achieved a high school diploma or a General Equivalency Diploma (GED) certificate should be eligible for adult literacy programs. Inmates who have achieved minimum scores for the issuance of a high school diploma or GED certificate (under the regulations of the state in which they are incarcerated) should be considered to have successfully completed an adult literacy program and should be awarded a high school diploma or GED certificate by the state.

English as a Second Language

English as a Second Language (ESL) is an educational program that offers English proficiency skill training at a basic functional level through advanced level of instruction. Inmates who score less than 225 points (eighth-grade proficiency) on the Comprehensive Adult Student Assessment System may be defined as having a need for ESL training. Those who take such training and are able to score 225 points or above may be considered to have completed the training successfully.

Occupational Education

Occupational education (OE), also known as vocational training, helps inmates acquire or improve marketable skills through one or more training programs. A specific inmate's need for OE may be determined through a review of his or her previous education and work history. When there is no demonstrated stable work history, or no specialized education or training record to demonstrate a marketable skill, the inmate will be judged to have a need for some form of intervention. Programs of occupational counseling, work experience, or formal training may be used, independently or in combination, to satisfy the identified need. Ordinarily, an inmate should not receive compensation for participation in an OE program. When compensation is received (for example, during apprenticeship training), such compensation must not exceed the amount appropriate for participation in work assignments (see "Inmate Employment," above).

An OE enrollment should be entered into the inmate's education record when he or she begins a program of study in any formal occupational skills training program. Final action will be entered into the record when the inmate exits that program. Successful completion will be certified when one or more of the following criteria are met:

- *Exploratory training:* programs designed to provide an introduction to a cluster of related occupations. An exploratory-level comple-

tion may be awarded when the inmate has met written criteria for the program that have been established by the institution and approved by the education administrator at the correctional agency's headquarters.
- *Marketable skill training:* achievement of marketable skills at the normal entry level for a specific occupational title or cluster of titles, and completion of at least 100 hours of program attendance. Concurrent academic education requirements (such as completion of the GED) may be established for marketable skill training programs. Inmates ordinarily should not be enrolled in any marketable skill training program if they have not previously met academic requirements or if they do not maintain concurrent enrollment.
- *Apprenticeship training:* achieving the Joint Apprenticeship Committee's requirements for a journeyman's certificate in a program registered under the U.S. Labor Department or an equivalent state registration program.
- *Postsecondary education activities:* obtaining a passing grade in a course approved for postsecondary credit by an accredited postsecondary education institution.
- *Adult continuing education (ACE) activities:* formal instructional classes in a special interest, such as typing, financial planning, parenting, refresher training in a basic skill, consumer education, computer literacy, etc. Classes in traditional leisure time activities, such as hobby crafts or music, should only be considered ACE activities if they have established curriculums and attendance that is both required and recorded. Successful completion of an ACE program occurs when the inmate has met the participation and attendance standards and the achievement standards that have been established for that activity.
- *Prerelease training:* an education program that assists the inmate with specific and broad-based preparation for release back into society. Successful completion occurs when the inmate has completed all course activities required in the policy of the pre-release program or by meeting established, agencywide standards.

- In addition, staff should encourage each inmate to accept the responsibility to identify any specific education needs, set personal goals, and select activities, programs, or work experiences that can help them reach those goals.

3. Academic and Career Counseling

- Institutions should provide academic and career counseling, which are self-help programs that can assist inmates with career planning and development. Such counseling should be an ongoing activity throughout an inmate's period of incarceration. Individual needs for specific types of assistance that can be addressed through career and academic counseling may be determined by considering a combination of factors, including educational level, work history, inventories of aptitude and interest, specific job skills, and unit team recommendations. Career and academic counseling should provide assistance, encouragement, and feedback regarding inmates' education and occupation programs while incarcerated and their postrelease goals. Individualized assessment programs should be included in the admission and orientation program when feasible (see Chapter 1).

4. Attendance and Records

- To be effective, institution education programs should monitor inmate attendance and maintain records of inmates' educational attainments. Teachers in education and vocational training classes should maintain current class roles and retain inactive class roles for at least one year. These roles can be used by unit teams in periodic reviews of an inmate's progress and program participation, and for administrative purposes by education staff at the institution and the correctional agency's headquarters.
- Attendance should be taken at classes and inmates should be expected to attend class regularly. The institution's education supervisor should monitor both excused and unexcused absences from classes and should be able to discern and correct patterns of interruptions in class attendance due to work details, sick calls, and other call-outs.
- An education file (in either hard copy or electronic format) should be maintained on each inmate. The file should include: (a) an interview record, documenting the inmate's education level and program interests; (b) documents relating to inmate participation in literacy and other education programs; (c) records of any instances where an inmate has been exempted from participation in a program, including the reasons for the exemptions; and (d) test scores, earned certificates and diplomas, etc. Education

files should be forwarded to the new institution when inmates are transferred or consolidated with the central file when inmates are released or sent to halfway houses.

5. General Program Characteristics

- Uniform standards should be followed in the operation of education and vocational training courses in all institutions that are part of the correctional agency. Uniform standards are necessary for purposes of program accountability and evaluation and for resource allocation. Further, clearly defined standards contribute to more effective program management and thereby can enlarge and improve inmate education and training opportunities.
- The Education Department at each institution should: (a) maintain written curriculums for each course, establishing measurable behavioral objectives and procedures; these curriculums should be approved by the education supervisor, followed by all classroom instructors, reviewed annually, and updated as necessary; (b) establish criteria to define minimum expectations for program completion, as well as requirements for the assessment of student progress, and permit inmates to enter or exit a program at their discretion at monthly intervals, except in unusual circumstances; and (c) meet all requirements set forth by the state Department of Education.
- Upon an inmate's successful completion of a program, the institution's Education Department should issue or file a certificate or other form of documentation verifying that successful completion, especially if such documentation can contribute to an inmate's future plans by validating his or her education and training, supporting an inmate's chances of securing postrelease employment, improving an inmate's prospects for being accepted into an advanced education program, or enhancing the inmate's opportunities for success in any legitimate activity he or she chooses to pursue.

Appropriate types of certificates and other documentation would include:
- accredited certificates, such as high school diplomas and occupational training certificates, which may be approved or issued through local high school districts, state departments of education, or other recognized accrediting educational organizations

- postsecondary certificates and transcripts, approved or issued through a sponsoring accredited educational institution
- GED certificates, issued for testing programs sponsored by the American Council on Education
- private certificates, approved or issued by private businesses or outside agencies
- institutional certificates, issued by the Education Department at the correctional facility for an inmate's successful completion of courses in general education, occupational training, adult continuing education, social education, etc., when certificates for such programs are not available from any of the organizations cited above
- transcripts, issued to inmates who have completed general education programs, formal occupational training, on-the-job and apprentice training, and work assignments.

With the inmate's consent, such transcripts may be sent to schools and colleges, potential employers, and other agencies.

- Bilingual and bicultural programs should be provided at any institution where there are significant numbers of non–English-speaking inmates. Necessary specialized equipment and facilities should be provided for all academic and occupational programs.
- Standardized competency-based curriculums must be supported by appropriate materials and classroom resources. Instructional objectives should be stated in terms that allow the performance to be observed or assessed. Emphasis should be placed on individual student progress and on evaluation based on the inmate's performance, according to established criteria. Instructors should maintain individual student progress charts that correspond to the competencies identified in the curriculum. Each institution should hold a graduation ceremony at least once a year to recognize inmates for program accomplishments, including successful completion of programs in ESL, GED, occupational training, postsecondary education, and other areas.
- Education and vocational training programs should be accredited by a state or other recognized accreditation association or agency. When there is no feasible method of accrediting the entire Education Department, special efforts should be made to acquire independent certification for each of the individual programs that are offered.

- A trade advisory committee or joint apprenticeship committee should be set up by the institution when occupational training programs are not offered by accredited education institutions or registered with the state's Department of Apprenticeship and Training or the comparable branch of the U.S. Department of Labor. When resources permit, accredited or registered occupational training programs are preferred. The size and composition of trade advisory committees will vary according to local needs but should include at least two active members who are not regular employees or contractors of the institution. The intent would be to include representatives from trade organizations, accredited training institutions, and potential employers. Such committees should meet at least twice a year.
- Education programs should operate on a twelve-month basis, with minimum break periods for holidays. Educational activities should be provided at least ten hours per day, Monday through Friday, although they do not need to be ten consecutive hours. For example, educational programs could be scheduled for 7–11 AM, 1–4 PM, and 6–9 PM. Weekend operation of education facilities is desirable wherever practicable.
- The institution's Education Department should develop an education handbook, providing an overview of all education programs, the incentives and achievement awards system, and other pertinent information. Education handbooks should be distributed to all inmates during the admission and orientation programs and should be produced in foreign language versions if significant numbers of inmates do not speak English. The handbook should be updated at least every two years.
- A member of the education or recreation staff should be assigned to serve as an education advisor on each unit team (see Chapter 1).

6. Staff and Volunteers

- Full-time instructors, whether they are regular employees of the correctional agency or contract employees, should spend at least 75 percent of their work week providing instruction or engaging in work related to instruction. The institution's education supervisor should ensure that annual formal classroom observations are conducted for all teachers, including written evaluations and recommendations for professional development.

- Institutions may establish programs utilizing qualified community volunteers and inmate tutors. Volunteers must be carefully screened and subjected to a check by the National Crime Information Center. Individuals who are approved inmate visitors may not serve as volunteers.

II. RECREATIONAL ACTIVITIES

Overview

In order to alleviate potentially disruptive inmate idleness, reduce personal stress and institutional tension, and promote positive lifestyles, the institution should encourage inmates to make constructive and socially responsible use of leisure time by offering movies, games, sports, social activities, arts and hobby crafts, wellness programs, and other group and individual activities. Such activities can increase physical wellness, thereby reducing inmate health care costs, and can contribute to personal and institutional stability by attaining maximum inmate participation in both formal and informal programs.

1. Inmate Needs

- All inmates have a need for informal recreation and leisure activities and should be able to participate in one or more leisure, fitness, wellness, or sport activities. The warden should ensure that the institution provides leisure activities designed to meet the social, physical, psychological, and overall activity needs of the inmate population.
- Certain inmates may benefit from more formal program involvement, due to medical, physical, or emotional needs. Such needs should be considered at initial classification meetings or when recommended by a member of the medical staff or the recreation staff.

2. Responsibilities

- The recreation supervisor should be part of the institution's Education Department, reporting to the education supervisor. Recreational activities should be developed by the recreation

supervisor in consultation with the education supervisor. In addition, the recreation supervisor should serve as a consultant to staff in all departments on matters regarding inmate physical fitness, wellness, and related areas.
- The recreation supervisor should draft an overview of institution recreational activities for inclusion in the education handbook (see Part I), which the institution distributes to inmates during the admission and orientation programs.

The recreation supervisor should also:

- develop a monthly activity schedule, which may be posted or published for inmate use (if appropriate)
- develop presentations for the admission and orientation program to provide information on leisure time activities and encourage participation
- prepare a recreation budget and perform other administrative tasks
- coordinate unit and institution recreation activities with unit management staff
- develop a general plan of action to identify "at-risk" inmates (i.e., those inmates who are reluctant to participate in recreational activities because they are overweight, have negative attitudes toward physical fitness programs, or are suffering from any mental, physical, emotional, or psychological handicaps or problems) and institute programs to encourage their participation in structured, unstructured, or spectator activities
- provide recreational activities at least eight hours a day, Monday through Friday, and at least twelve hours a day on weekends (in cases where significant numbers of inmates do not have daily access to recreational activities during regular hours because of work assignment conflicts, the recreation supervisor should arrange for special weekday morning recreation supervision for those inmates and should ensure that they have priority access to hobby crafts and music facilities on weekends)
- ensure that all necessary specialized equipment and facilities are available for recreational and leisure-time programs
- develop guidelines for the use of protective equipment and clothing and implement procedures to disseminate and enforce these guidelines
- oversee registration and the maintenance of rosters for organized activities

3. Movies and Television

- If the institution shows movies to inmates, the recreation supervisor should exercise good judgment when selecting video rentals, particularly with respect to sexual and violent content. In particular, he or she should ensure that X-rated and NC–17–rated films are not featured. In addition, the warden may wish to prohibit movies focusing on criminal activities and prisons.
- The use of videotapes in public performances (including showings to groups of inmates) is prohibited without a licensing agreement. All videotape copyright and licensing requirements must be strictly enforced.
- Unless specifically authorized by statute, institutions may not purchase premium movie channels or video licenses for inmate viewing. With the approval of the appropriate executive-level official at the correctional agency's headquarters, the institution may purchase basic cable service if local open-air television programming is not available.
- Inmate commissary funds may not be used to fund cable television service. At the warden's discretion, however, a specialty television channel or a video license and rentals may be purchased with inmate organization funds (see Part IV).

4. Athletic Activities

- Traditionally, inmates have been permitted to participate in such athletic activities as weightlifting, soccer, baseball, softball, basketball, running, and, at some institutions, tennis. For several reasons, including security concerns and public affairs considerations, laws or regulations in many states have prohibited some of these activities—in particular, weight training using free weights. Inmate participation in such sports as boxing, wrestling, martial arts, etc., should be prohibited. Recreation supervisors should be alert to all laws and regulations in their jurisdictions affecting inmate athletic pursuits and should be prepared to promote approved alternative athletic activities—such as calisthenics, aerobics, or isometrics—if other popular activities are prohibited.
- Running events ordinarily should not exceed 10 kilometers, or 6.2 miles. Appropriate medical staff and fluids (waters, sports drinks, etc.) should be available for all inmate running events.

- When all members of an inmate sports team are at an appropriate custody level, the team may participate—under escort—in sports activities outside the institution. In addition, the warden may authorize appropriately screened sports teams from the community to visit the institution to play inmate teams.

5. Arts and Hobby Crafts

- Art work would include all paintings and sketches rendered in any of the usual media (such as oil paints, pastels, crayons, pencils, inks, and charcoal). Hobby craft activities would include ceramics, leatherwork, model making, clay, mosaics, crochet, knitting, sculptures, woodworking, lapidary, and other pursuits that are consistent with institution guidelines. At a minimum, musical activities would involve the use of stringed instruments. Electronic instruments, including keyboards and electronic drums, may also be permitted, unless prohibited by law or agency regulations. Brass instruments, standard drums, pianos, etc., may be permitted if, for example, the institution has an inmate band.
- Art, musical, and hobby craft tools, instruments, and supplies may be obtained through the institution's recreation program, the commissary sales unit, special purchase commissary orders, or other sources approved by the warden.
- Each inmate must identify completed art or hobby craft items by marking the reverse side of the item with his or her name and registration number. Completed or abandoned art or hobby craft items should be disposed of by giving the item to an authorized visitor (with the approval of the warden), mailing the item to a relative or approved visitor (at the inmate's expense), or selling the item through an institution art and hobby craft sales program (if one exists), after the institution price committee has determined the sale price. It must be noted, however, that art and hobby craft programs are intended for the personal enjoyment of the inmate and as an opportunity to learn a new leisure skill; although individual items may be sold, the art and hobby craft programs are not intended to facilitate the mass production of such items or to provide a significant means of supplementing the inmate's income.
- If art and hobby craft items are not disposed of in any of the ways cited above, and if the inmate does not have enough space to keep

such items in his or her living area, the items should be removed and disposed of at the warden's discretion. Restricting the volume of art and hobby craft items within the institution is necessary to reduce fire hazards and conserve space.
- The warden may restrict, for reasons of fire safety, security, and housekeeping, the size and quantity of products made in the art and hobby craft programs and the use or possession of art or hobby craft items or materials.
- All hazardous materials should be stored and disposed of in accordance with institution safety regulations and regulations issued by OSHA. All such materials should be inventoried regularly. Material safety data sheets should be maintained for all toxic, flammable, and caustic materials and should be available for all staff and inmates to consult. All persons involved with art or handicraft materials that are hazardous should receive safety instruction in accordance with the *Occupational Safety and Environmental Health Manual*.
- The warden may limit hobby craft and art projects in an inmate's cell or living area to those that can be stored in the personal property containers provided by the institution. Exceptions may be made for such items as paintings that are too large to be placed in a locker. Art and hobby craft items must be removed from the living area after they have been completed, unless they are approved as personal property.
- If an inmate's art work or hobby craft product is to be put on public display, the warden may restrict the content of the work in accordance with community standards of decency. To ensure that legally appropriate standards are met, the warden should consult with the correctional agency's legal counsel before removing or restricting any particular item.
- Where space and equipment are limited and demand is high, the warden may set limits on the amount of time an inmate may use a hobby craft facility. For example, the warden may limit an inmate's use of any workshop or classroom to six months, or to four hours a week, in order to make room for other inmates who wish to work on a project. Hobby craft and art participants may be permitted to use facilities and materials on a rotating basis to allow for maximum utilization of resources.
- Disciplinary action may be taken against any inmate found with unauthorized art or hobby craft materials in his or her possession. One sanction may be to remove the inmate from the art or hobby craft program.

6. Other Activities

- In consultation with the Health Services Department, the recreation supervisor should establish an inmate wellness program, to include screening, assessments, goal setting, fitness and nutrition prescriptions, and counseling.
- A supply of appropriate board games (chess, dominoes, etc.) should be available for inmates. Working with the education supervisor, the recreation supervisor should make sure that the inmate academic library includes a section of appropriate titles for leisure reading. Such volumes may be acquired through library purchases or through donations.

III. RELIGIOUS ACTIVITIES

Overview

Each institution is obligated to provide inmates of all religious faiths with reasonable and equitable opportunities to pursue individual religious beliefs and practices, within the constraints of budgetary limitations and the need to operate the institution in an orderly and secure manner. Under the national Religious Freedom Restoration Act (RFRA), new and changing obligations have been placed on prison systems; consequently, institution staff should consult the correctional agency's legal counsel for the latest judicial rulings on RFRA requirements.

1. Procedures and Guidelines

- Institution chaplains should be available upon request to provide pastoral care and counseling. They should schedule and coordinate all religious activities in the institution. The warden may be responsible for performing the administrative or organizational functions of a chaplain if the institution does not have a full-time chaplain on staff.
- The institution must provide space and equipment adequate for the conduct and administration of religious programs. In the scheduling of approved religious activities, both the availability of staff supervision and the need to share available time and space

proportionately among eligible groups should be taken into consideration.
- Depending on the size and religious needs of the inmate population, the institution may hire one or more chaplains, representing different religious faiths, to provide spiritual guidance and services. In addition, to help accommodate inmates from as many faith groups as possible, and help them pursue their religious beliefs, the institution may hire part-time chaplains on a contract basis or accept the services of volunteers from the community. The representatives of faith groups who may be hired, or whose services as volunteers may be accepted, would include both clergy and lay spiritual advisors. The chief chaplain or the warden may require a recognized representative of the faith group to verify the status of a religious representative before approving that person's entry into the institution as either a contract chaplain or religious volunteer.
- No one may disparage the religious beliefs of an inmate, or coerce or harass an inmate to change religious affiliation. Staff are prohibited from proselytizing inmates.
- Attendance at all religious activities is voluntary, and no inmate may be required to profess a religious belief. An inmate may designate any religious preference or may designate no religious preference; he or she may change this designation at any time.
- It is conceivable that an inmate could show a pattern of changing religious preference immediately preceding the occurrence of a faith group's scheduled religious activity that he or she may perceive as providing special advantages. To prevent this specific abuse, the warden—in consultation with the chief chaplain—may exclude an inmate from participating in that activity. In such cases, a report explaining the reasons for the exclusion should be placed in the inmate's central file.
- An inmate ordinarily should be allowed to wear or use personal religious items during devotional services. This includes robes, prayer shawls, prayer rugs, medicine pouches, religious headbands, kufis, yarmulkes, beads, phylacteries, and medallions. Upon specific request of the inmate, the warden may also allow an inmate to wear or use personal religious items anywhere in the institution. The warden's decision on whether an inmate may wear or use personal religious items, however—whether during devotional services or at other times—should be based on considerations of security, safety, and good order. Even if the warden determines that wearing or using a particular religious item through-

out the institution could violate such considerations, he or she may nonetheless permit the inmate to wear or use it during devotional services.

The warden may request the chief chaplain to obtain a determination from representatives of the inmate's religious faith group or other appropriate sources concerning the religious significance of any items in question. Secure storage space should be provided for any religious items that may be worn or used during religious services but might compromise institutional security, safety, or good order at other times.

- Each inmate who wishes to have religious books, publications, or other materials must comply with institution guidelines regarding ordering, purchasing, retaining, and accumulating publications and other personal property (see Chapters 1 and 3).
- A reasonable portion of the budget for chaplaincy services may be used to acquire religious literature for inmate use. Acquisition of such literature should be carried out in an equitable fashion, reflecting the diversity of religious faith groups represented in the inmate population.

2. Religious Diets

- The institution must provide those inmates who desire a religious diet with reasonable and equitable opportunities to observe religious dietary laws, within the constraints of budgetary limitations and the need to operate the institution in a secure and orderly manner.
- An inmate who wishes to observe religious dietary laws should be given a diet that meets or exceeds recommended daily allowances established by the Food and Nutrition Board of the National Research Council, National Academy of Sciences, and that complies with religious dietary laws to the best extent practicable
- As a once-a-year accommodation, the institution may arrange for an inmate religious group to have a ceremonial meal, upon the specific request of inmates in that group. Staff may purchase from a supplier specially prepared food items that meet religious requirements. In making its request, the inmate religious group should provide information on the religious significance of the food items. In addition, the warden may request the chief chap-

lain to obtain a determination from representatives of the faith group or other sources regarding the religious significance of the items in question. Funds for the purchase of special food items for a ceremonial meal may be provided from the budget for chaplaincy services, from the institution's Food Service Department, or from community organizations.

In addition, the warden may excuse inmates from work for one day, in connection with the annual ceremonial meal (see "Religious Holidays, Celebrations, and Activities," below).

3. Institution Work Assignments

- An inmate may not be given a work assignment that violates the specific requirement of his or her religious faith, except where necessary for maintaining safety, security, and good order within the institution. For example, an inmate whose religious teachings prohibit the handling of pork should not be assigned to a position requiring the inmate to prepare, serve, or handle pork or pork products. Any request for a different work assignment based on religious concerns must be initiated by the inmate.

In cases where a change in work assignment has been requested, the chief chaplain or other appropriate religious consultant should verify the specific religious requirements in question.

4. Religious Holidays, Celebrations, and Activities

- The warden should facilitate the observance of important religious holidays or celebrations in accordance with specific requirements of a faith group, such as fasting, worship, diet, or work proscription, provided that it is consistent with maintaining the security, safety, and good order of the institution.

The inmate must initiate the request for observations of a religious holiday or celebration. The nature of the observance and the particular accommodation requested must be specified. The warden may ask the chief chaplain to verify the specified religious requirements with representatives of the inmate's faith group or other appropriate sources.

Ordinarily, the warden should allow an inmate to take nonconsecutive vacation days, make up missed work, or change work assignments in order to facilitate the inmate's request for observance of a religious holiday or celebration.

- The warden may excuse an inmate from an institution program or assignment if a religious activity is also scheduled at that time. The more central the particular religious activity is to the tenets of the inmate's religious faith, the greater the presumption is for excusing the inmate form the institution program or assignment. To help determine the appropriateness of excusing an inmate in such instances, the warden may consult with the chief chaplain regarding the faith group's religious tenets.

IV. INMATE ORGANIZATIONS

Overview

With the warden's approval, inmates may establish formal organizations for recreational, social, civic, and benevolent purposes. Such inmate organizations may include individuals from the outside community. Inmates are eligible to participate only in those organizations that have been recognized and approved by the warden.

1. Requirements

Any proposed inmate organization must submit a request for recognition to the warden before it may become active. The warden may approve an inmate organization when it meets all the following criteria:

- The organization must not operate in any way that would compromise the security, good order, or discipline of the institution.
- The organization must have a constitution and by-laws that state its purpose and operations and also set forth the duties and responsibilities of the officers. The warden may amend the constitution and by-laws at his or her discretion.
- The organization must be coordinated by a staff sponsor, whose duties should be performed while on official duty status. In addition, staff members may volunteer to work with inmate

organizations during their off-duty time, but this would not be part of the staff sponsor's official responsibilities.
- All meetings of the organization must be approved by the warden and supervised by staff members. The organization may not hold meetings at times that compete with scheduled inmate work and program activities.

2. Fund Raising, Expenditures, and Financial Accountability

- Inmate organizations may collect dues from members if the warden has approved the rate and method of collection, but they cannot make the payment of dues a requirement of membership for any inmate who lacks funds.
- All activities and projects sponsored by the organization require the approval of the warden, including fund-raising projects. Inmates should do most of the work on fund-raising projects for inmate organizations. The warden may not approve a fund-raising project that competes against the commissary, creates unreasonable additional work for staff, requires supervision beyond the resources available to the institution, or interferes in any way with the orderly operation of the institution. No fund-raising activity or program may be connected to gambling.
- On occasion, an inmate organization may request to sell tickets as part of a fund-raising effort (including coupons or tokens that may be exchanged for food items or admission to a special showing of a movie, for instance). Before authorizing an inmate organization to sell such tickets, the warden must require the organization to establish an accountability procedure to ensure that it has sufficient reserves to redeem all outstanding tickets. Further, to ensure that such tickets are redeemed, and to prevent such tickets from being hoarded by inmates as a medium of exchange, the design or color of the tickets should be changed periodically or expiration dates should be written on them.
- Upon approval of the warden, an inmate organization may use its funds to provide financial assistance (including loans or grants) to inmates with insufficient resources, in order to defray expenses for an emergency humanitarian purpose, such as a bedside visit to a critically ill relative or a funeral trip. The warden may authorize a separate fund (an inmate welfare fund, inmate emergency fund, etc.) to be created for this purpose. Operating methods of any such

undertaking should be clearly established, indicating criteria that an inmate must meet to receive a loan or grant, the terms of loans (including repayment), the source of funds, any restrictions that exist on the use of the funds, and the method of accountability for these funds. Such funds should be maintained exclusively by institution staff, but inmates' representation is required when the use of funds is being determined.
- Upon approval of the warden, an inmate organization may be allowed to purchase items for use in the institution, provided that such items are purchased *in addition* to items normally furnished by the government (i.e., the government may not rely on inmate organizations to provide funds for items that the government should provide).

For example, an inmate organization may use its funds to purchase a television, a videotape recorder, games, toys, or babysitting services for use in the institution visiting room. An inmate organization may also be permitted to purchase supplemental items for inclusion in the annual holiday package distributed to inmates or to hire entertainers for special events.

- Every inmate organization should appoint an inmate treasurer to keep financial records, in accordance with generally accepted accounting procedures, reflecting income identified by source and expenditures with applicable receipts. Whenever possible, expenditures must be in the form of checks, disbursements must be approved in accordance with the organization's constitution, and any expenditures for travel must be undertaken in accordance with the government's travel regulations and documented with appropriate vouchers. Checking accounts established by inmate organizations should require at least two signatures for a check to be issued, and one must belong to the staff sponsor or a deputy warden.

The treasurer should prepare financial reports immediately following the end of every fiscal quarter and present copies to the membership, the staff sponsor, the institution business manager, and the warden. In addition, the warden should require a financial audit of each inmate organization immediately following the end of each fiscal year. If possible, the audit should be conducted by a commercial accounting firm, at the organization's expense. If the organization does not have sufficient funds to hire a commercial accounting firm, the warden may assign responsibility for conducting the audit to a staff member who has accounting experience but is not associated with the organization. All audits should be reviewed by the institution's business manager, who should report deficiencies or recommendations to the warden.

The warden should disband any inmate organization that fails to maintain suitable financial records, furnish financial statements, or complete the required annual audit.

3. Special Activities

- Banquets, community programs, charitable contributions, or the attendance of guests from outside the institution at meetings and activities of the organization require the warden's approval. Normally, the warden should require guests to purchase a meal ticket when attending banquets where the government incurs the costs. Meals may be furnished without charge, however, to persons rendering a special service to the institution and certain other visitors, in accordance with agency policies permitting specific exemptions from the requirement to purchase meal tickets.
- All special activities should be under the direct supervision of the inmate organization's staff sponsor or coordinator.

V. HALFWAY HOUSE PLACEMENT

Overview

Halfway house programs provide an excellent transitional environment for inmates nearing the end of their sentences. The level of structure and supervision ensures accountability and provides opportunities in employment counseling and placement, substance abuse, and daily life skills. While dangerous inmates should be separated from the community until the completion of their sentences, inmates who are eligible for halfway house placement should be referred to halfway houses in order to maximize their chances for reintegrating into society. Referring inmates to halfway houses can enhance public safety by helping the offender make a successful transition back into the community. Participation in community-based transitional services reduces the likelihood of recidivism by inmates with limited resources.

Institution staff are not involved in the operation of halfway houses, which often are contract facilities. They do play a critical role, however, in preparing an inmate's release plan, determining inmate eligibility for halfway house placement, applying criteria and guidelines, and carrying out procedures relating to an inmate's transfer to a halfway house.

1. Overview of Halfway House Programs and Community Corrections

- Community corrections generally refers to a range of sentencing or designation options that do not involve incarceration in secure correctional facilities. Halfway houses are the most common component of the community corrections matrix. Typically owned and operated by private or nonprofit contractors, halfway houses provide supervised living quarters for offenders in the final months of their sentences and generally provide other services (including drug testing, drug and alcohol counseling, mental health counseling, job placement, and other program activities). Inmates serving in halfway houses may be physically restricted to the halfway house, except for employment or other structured program activities, or may be in a less restrictive prerelease program, under which they have greater access to family members and the community and may receive weekend and evening passes.
- Other types of community corrections options include home confinement at the offender's residence (usually involving use of electronic monitoring devices); urban work programs (whereby offenders are confined at a halfway house for a considerable portion of their sentences, are assigned to work on government-sponsored projects, and are under the more restrictive form of halfway house confinement until becoming eligible for the prerelease program during their last months of their sentences); comprehensive sanctions centers (facilities that offer a wider range of sentencing options, programs, treatment and training opportunities, and levels of supervision than traditional halfway houses and are geared toward providing prerelease services for higher-risk inmates); and intensive confinement centers (which usually are adjacent to and operated by a correctional facility, are similar to satellite camps, and are commonly known as "boot camps" because they involve military-like training and regimens).

- In many cases, the different components of the community corrections matrix are designed to operate in conjunction with each other. For example, assignment of an offender to a home confinement program frequently occurs after that offender has served in a halfway house. Similarly, intensive confinement involves both the successful completion of the program in the intensive confinement center itself, as well as mandatory placement in a halfway house thereafter.
- Offenders who are employed may be required to pay a subsistence charge to defray the costs of their participation in a community corrections program (particularly the halfway house or home confinement phases).

2. Release Plan

- The inmate's unit team should begin release planning at its first meeting, which usually would be the initial classification meeting. Release planning should continue throughout the inmate's confinement (especially during regular unit team meetings to review the inmate's progress).
- By planning for the inmate's release early in the inmate's period of confinement, the unit team can ensure that prerelease needs are identified and appropriate prerelease programs are recommended. Preliminary decisions regarding an inmate's eligibility for halfway house placement should be made well before the inmate begins his or her last year of confinement. The unit team should develop a final and specific prerelease plan, including a decision regarding halfway house referral, no later than eleven months before an inmate's projected release date.

3. Halfway House Criteria, Eligibility, and Referral Guidelines

- The goal of halfway house placement is to maximize each eligible inmate's chances for a successful reintegration into the community and his or her prospects for leading a law-abiding life after release. Halfway house placement is a program element and must not be used as a reward for good institutional behavior. Nevertheless, an inmate's institutional adjustment may be a factor in making a referral determination.

- Staff must weigh several factors in determining whether an inmate will be referred for halfway house placement, and how long the placement should be. Those factors include the inmate's needs for services, the availability of community resources, and concerns of public safety. To manage bed space usage effectively, institution staff should make a recommendation for a specific placement date to the community programs manager, and halfway house staff should determine actual placement dates and monitor average lengths of stay. Budgetary constraints are a key consideration and may affect average lengths of stay. Halfway house staff should make every effort to establish an acceptable date as close to the institution's recommended date as possible. The actual date of an inmate's placement in a halfway house, however, should never precede the institution's recommended date.
- State law or agency regulations may establish a maximum number of days for which inmates may be placed in halfway houses, pending release. Placement beyond that period should be permitted only in extraordinary cases, with the approval of the director or commissioner of the correctional agency, the supervising probation or parole officer in the inmate's sentencing jurisdiction, and the sentencing judge.
- Inmates with a history of escape from or failure in one or more halfway house programs should be subject to extremely careful review regarding their suitability for placement and the appropriate length of placement. Inmates with physical handicaps, minor medical problems, or disabilities should be considered for personal placement on the same basis as any other inmate. It should be noted, however, that inmates in halfway houses are required to assume financial responsibility for their own health care; those unable or unwilling to do so may be denied placement.

Inmates in the following categories ordinarily should not be placed in halfway house programs:
- sex offenders whose current offense or behavioral history indicates involvement in predatory or assaultive sexual behavior
- deportable aliens
- inmates who require mental health treatment or inpatient medical care (this category does not include inmates participating in nonresidential drug abuse treatment through psychology services, or who are utilizing mental health services for adjustment or other psychological problems that do not have serious implica-

tions affecting an inmate's ability to benefit from halfway house placement)
- inmates who refuse to participate in a financial responsibility program established by the agency
- inmates with pending charges or detainers
- inmates serving sentences of six months or less (not including inmates who have been sentenced by the court directly to a halfway house)
- inmates who pose a significant threat to the community, as determined by repeated or serious institution rule violations, a history of repetitive violence, escape, association with gangs or other types of violent or terrorist organizations, or other factors reflecting an inmate's proclivity for violence or escape.

4. Inmate Refusals of Halfway House Placement

- There may be occasions when an inmate who is eligible for halfway house placement refuses to accept such placement. Staff members should investigate the inmate's reasons for declining placement. Appropriate reasons for an inmate to refuse placement in a halfway house would include previous halfway house failure, potential conflict with other residents of the halfway house, or the location of the halfway house (especially if it is remote from the inmate's intended residence after release).
- In most cases, the institution should honor an inmate's refusal of halfway house placement. If the inmate fails to present appropriate reasons for declining placement, and the unit team believes that placement would serve a valid correctional or agency need, then the unit team should make every effort to encourage participation.
- When an inmate refuses placement, a memorandum should be signed by the deputy warden and the inmate documenting the inmate's rationale for refusal and citing any efforts by the unit team to encourage participation. The memorandum should be placed in the inmate's central file.

5. Halfway House Referral Procedures

- At least eleven months before an inmate's probable release date, the unit team should decide whether to refer the inmate to a halfway house or other community program. If the unit team decides to recommend a transfer, it should complete the appropri-

ate referral form and forward it to the warden for a final decision. If the warden approves the recommendation, the unit team should forward two copies of the referral form (with appropriate attachments) to the halfway house and the community programs manager, at least sixty days before the requested placement date. If the placement is for forty-five days or less, and the release method is by parole, a copy of the correspondence with the parole board that outlines the release plan and requests parole certificates, as well as copies of the parole board's response recommending release plan approval, should be included in the referral package.
- Ordinarily, halfway house staff may not refuse to accept a referral. If a rejection does occur, halfway house staff should provide specific reasons for doing so, in writing. The warden should then determine if further discussion of the case with the halfway house staff is appropriate, or if referral to an alternate resource is possible.
- If an inmate cannot be transferred to the halfway house on the scheduled date, the institution staff should notify the halfway house staff immediately.
- At least three weeks before the approved transfer date, the unit staff should determine how much money from the inmate's commissary account may be given to the inmate. A check or draft for the approved amount, with the inmate as payee, should be sent to the halfway house immediately. The unit staff may recommend payment of a gratuity to those inmates who lack resources or funds.

The unit staff should use discretion in giving inmates large amounts of cash before their actual release. If there is any reason to question an inmate's ability to handle money responsibly, then the amount may be less than the balance in the account (this does not affect the inmate's ownership of the money in his or her commissary account, but only means that the balance may not be provided to the inmate until some later point in the inmate's confinement in the halfway house, or until the inmate's release).

If an institution is holding savings bonds for an inmate, or if an inmate has a savings account at a local bank, the unit staff should ensure that those financial resources are available to the inmate at the release destination, when he or she arrives there.

- At least two weeks before an inmate's approved transfer date, institution staff should forward the following documents to the halfway house:

— the Authorized Unescorted Commitment and Transfer form, with current photograph
— the original transfer order
— a copy of the Furlough Application and Approval Form, with specific travel method and itinerary
— a receipt for halfway house rules and regulations and the Subsistence Agreement Form, signed by the inmate.
- At least one week before the inmate's approved transfer date, arrangements should be made to provide the inmate with release clothing, including attire that would be appropriate for carrying out a job search and performing work, and an outer garment that would be appropriate for weather conditions at the inmate's release destination. In addition, the institution staff should assist the inmate in acquiring appropriate identification (such as a driver's license and birth certificate), the inmate should be assigned to community custody status, final action should be taken by the parole board on any disciplinary reports that might affect a parole date, and the records manager of the sending institution should update the inmate's sentence computation.

Also no later than one week before the transfer date, the medical staff should review the inmate's medical record to determine if the inmate is on continuous medication; if so, the inmate should be provided with a thirty-day supply of medication. (If there is a possibility of abusing the medication, only a one-week supply should be given to the inmate, with the supply for the remaining three weeks being sent directly to the halfway house director on the day of the inmate's departure from the institution).

- If transfer dates are not established in time for the staff to carry out all the procedures cited above, then those procedures should be carried out as soon as possible after the transfer has taken place. Notifications of transfer to and arrival at the halfway house should be carried out as outlined in Chapter 2, Part XII.

VI. FURLOUGHS

Overview

A furlough is an authorized absence from an institution by an inmate who is not under escort by a staff member or other law enforcement officer. When granted to inmates under rigidly prescribed conditions, a furlough is a privilege granted to help inmates attain correctional goals. Furloughs are not a right, nor may they be given as a reward for good behavior or as a means to shorten a criminal sentence. A furlough may be authorized within the geographic boundaries of the state, or some smaller area within the state. Overnight furloughs ordinarily would be from three to seven calendar days long; lengthier furloughs may be authorized for medical, educational, or vocational needs, if permitted by law. Violations of furloughs may result in internal disciplinary action or even criminal prosecution against the inmate.

1. Duration of Furloughs

- With regard to duration, there are two types of furloughs—day furloughs and overnight furloughs.
- A day furlough may occur within the geographic limits of the commuting area of the institution (i.e., a radius of approximately 100 miles), may last no more than sixteen hours, and must end before midnight. Day furloughs typically are used to strengthen family ties and to enrich specific institution program experiences. Such furloughs frequently are associated with program activities (religious, educational, recreational, etc.) or activities of inmate organizations (Jaycees, Toastmasters, etc.) that take place outside the institution.
- An overnight furlough is one that falls outside or beyond any of the criteria of a day furlough, and generally will be authorized for three to seven calendar days.

2. Expenses of Furloughs

- Except in such instances as noted below, inmates, their families, or other appropriate sources approved by the warden are responsible for all expenses of a furlough, including transportation, food, lodging, and incidentals. The government may bear the expense of a furlough only when the purpose of the furlough is to obtain necessary medical, surgical, mental health, or dental treatment that is not otherwise available; to transfer an inmate to another correctional institution or halfway house; or to participate in some activity or assignment that is considered to

benefit the state (such as a special industrial or institutional work assignment).
- Institutions should pay transportation costs for inmates transferring to halfway houses, but, in such cases, inmates must travel directly from the institution to the halfway house and must use the mode of transportation directed by the institution. If an inmate wishes to deviate from the furlough travel schedule or take a different mode of transportation than is authorized by the institution, he or she must bear all transportation expenses.

3. Justifications for Furloughs

Furloughs may be authorized for the following reasons:
— for the inmate to be present during a crisis in his or her immediate family, or in other urgent situations
— for the inmate to participate in the development of release plans
— to enable the inmate to reestablish family and community ties
— for the inmate to participate in selected and approved educational, social, civic, religious, and recreational activities
— for the inmate to transfer directly to another correctional facility or halfway house, provided that the inmate is a voluntary surrender holdover case en route to a minimum-security facility or is a minimum-security inmate transferring either to another minimum-security facility or to a community corrections facility or program
— for the inmate to receive treatment at a medical referral center, provided that the inmate is in the lowest custody category, has been determined by the warden to be physically and mentally capable of completing a furlough, and has demonstrated sufficient responsibility to provide reasonable assurance that he or she would meet furlough requirements
— for the inmate to appear in court (whether in connection with a civil action, in order to comply with an official request to appear before a grand jury, or to appear in a criminal court proceeding in instances when the use of a furlough has been requested or recommended by the court or the prosecuting attorney)
— for the inmate to comply with official requests to appear before a legislative body, regulatory agency, or licensing agency

—to enable the inmate to participate in special training courses or institutional or industrial work assignments, including work assignments of thirty days or less in cases where daily commuting from the institution is not feasible
- Questions about the specific applications of the guidelines cited above should be referred to the correctional agency's legal counsel. Furloughs for purposes other than those specified in the guidelines may, at the discretion of the warden, be referred to the appropriate executive-level official at the correctional agency (i.e., a deputy director or deputy commissioner) for review and possible approval.

4. Eligibility

- Furloughs may be granted only to inmates with community custody. The one exception to this would be for inmates without custody, when the purpose of the furlough is to transfer directly to another institution or to obtain medical treatment.
- The inmate must be physically and mentally capable of completing the furlough, in the opinion of the warden. The inmate must have demonstrated sufficient responsibility to provide reasonable assurance that he or she will meet furlough requirements, in the opinion of the warden.
- Emergency furloughs and day furloughs may be granted only to inmates with no more than two years remaining before their anticipated release date (i.e., their mandatory or statutory release date, with good time credits; their full service of sentence date; or their parole date—whichever is earlier). Overnight furloughs inside the institution's commuting area may be granted only to inmates with no more than eighteen months remaining before their anticipated release date. Overnight furloughs outside the institution's commuting area may be granted only to inmates with no more than one year remaining before their anticipated release date.
- Furloughs ordinarily should not be granted to an inmate convicted of a serious crime against a person, or whose presence in the community could attract undue public attention, create unusual concern, or depreciate the seriousness of the offense. In cases where such an inmate is approved for a furlough, the reasons for the furlough should be documented in the inmate's central file.

Furloughs ordinarily should not be granted to an inmate who refuses to participate in the financial responsibility program (if one has been established by the institution or the correctional agency), who was determined under the institution disciplinary process to have used drugs or alcohol in the institution within the previous two years, who has a prior history of escape or attempted escape, or who has a prior history of committing sexual assault. Furloughs for such inmates should be granted only under highly unusual circumstances and with the written approval of the appropriate deputy director or deputy commissioner at the correctional agency's headquarters.

- Inmates under special monitoring status may be considered for furloughs only under the provisions of the policy governing special monitoring cases. The correctional agency does not have the authority to furlough unsentenced prisoners, and any requests for furloughs of such prisoners should be referred to the appropriate court of law.

5. Procedures

- Authority to approve furloughs must not be delegated below the level of warden or acting warden. Medical furloughs must be approved by the medical services administrator at the correctional agency's headquarters, upon the recommendation of *both* the warden and the institution's chief medical officer.
- An inmate who meets the eligibility requirements for a furlough may initiate the process by submitting an application for a furlough to his or her unit team for review. An inmate requesting a furlough ordinarily should undergo an HIV test; inmates refusing to do so may not be considered for a furlough. As part of the review to determine if a furlough should be recommended, the unit staff should consult with institution medical staff regarding the inmate's HIV status and general physical and mental ability to complete the furlough.
- The unit team should determine the validity of the furlough request by consulting institution medical staff (if the request is for a medical furlough), staff in the appropriate program area (if the furlough is to participate in religious, educational, work-related, or other program activity outside the institution), or by contacting family members or other persons that the inmate intends to visit during the furlough.

- If the furlough request is the inmate's first, then the unit team should send a questionnaire regarding the inmate's suitability for a furlough to the probation or parole office in the sentencing jurisdiction. If the furlough is to take place in a jurisdiction other than the sentencing jurisdiction, a questionnaire should also be sent to the probation or parole office in that jurisdiction (provided the furlough is the first to be made by that inmate to that particular jurisdiction). If the questionnaire is not returned within two weeks, the unit team should contact the probation or parole office to determine the status of the request. If the questionnaire is not returned within one week thereafter, staff should proceed with their processing of the furlough request.
- The warden may grant a furlough even if the appropriate probation or parole office opposes it, but he or she should document the reasons for approving the furlough, should notify the probation or parole office in writing, and should place copies of all documentation in the inmate's central file.
- If the warden and the probation or parole officer in the local jurisdiction concur, a blanket approval memorandum for all furloughs or for specific types of furloughs may be used in lieu of the prescribed questionnaire. Any blanket approval memorandum would require the signatures of both the warden and the chief probation or parole officer for that jurisdiction and should be reviewed and re-signed by both parties every other year.
- The warden should never approve a furlough to which an objection has been raised by a judge in the sentencing court or in the district to be visited.
- Before being routed to the warden, the furlough request should be submitted to the records office for a final detainer check and legal status check. If the warden approves the furlough, he or she should sign the original and at least three copies of the furlough request form. If the warden approves a furlough that does not fall under the guidelines cited in "Justifications for Furloughs," above, the reasons for doing so should be documented in the inmate's central file.
- Separate requests should be submitted for each furlough. Exemptions may be granted by the deputy director or deputy commissioner in charge of correctional management in cases where a furlough is needed for an extended period of time on a recurring basis.
- The staff should notify the inmate when a decision on his or her request for a furlough has been made. If the furlough is denied,

the staff should furnish the inmate with the reasons for the denial. If the furlough is approved, the staff should ensure that the inmate's mode of transportation is the same as listed on the furlough form. An inmate who is approved for a furlough must abide by the specified conditions of the furlough. In addition to general conditions applied to all furloughs, the warden may establish additional furlough conditions, as warranted.
- A correctional counselor should interview all inmates returning from furloughs to determine if furlough conditions were met. If anything unusual occurred during a furlough, the correctional counselor should document it in the inmate's central file, should notify the unit team, and should contact the probation or parole officer, the medical facility (if the furlough was for medical purposes), the inmate's family, or other appropriate resources for further information.

6. Violation of Furlough Conditions

- An inmate who absconds from a furlough should be considered an escapee. Escapes, or any other kind of serious incident or violation of furlough conditions, should be reported immediately to the security administrator and case management administrator at the correctional agency's headquarters.
- Disciplinary action may be initiated against any inmate who fails to comply with any of the conditions of the furlough, and, if deemed appropriate by staff, the matter may be referred to other law enforcement agencies for investigation and possible prosecution (see Chapter 2, Part VIII).

7. Public Information

- Institution and headquarters staff assigned to public information duties should promote public understanding of the furlough program by explaining to the public how the program relates to the total correctional process. Pertinent segments of local communities, such as civic organizations, should be kept apprised of any modifications to the program.

Chapter 5

Medical and Psychological Issues

I. INFECTIOUS DISEASE MANAGEMENT

Overview

Contagious disease processes are always a matter of concern in any environment, but they are particularly dangerous in the confined environment of a prison. They can have a devastating impact in any correctional facility and can place severe pressures on an institution's financial and human resources. Contagious disease processes or conditions, such as hepatitis, tuberculosis, and HIV infection, can have catastrophic results for the individual patient, can spread quickly to other inmates and staff, and can pose difficult management problems. Each institution should have procedures in place for the prevention and treatment of contagious diseases, and, as a last resort, the confinement of inmates suffering from contagious diseases in special housing situations, in order to help treat their diseases and to protect the health and safety of staff members and other inmates.

1. Health Promotion Initiatives

Disease prevention programs have been used effectively to limit the impact of communicable diseases and chronic medical conditions. Diseases and conditions detected in the early stages often may be cured or the disease process arrested. Effective health promotion and disease prevention programs emphasize the individual as a cooperative participant in the health care process rather than a passive recipient. The health promotion concept encompasses the following:

- Disease prevention, which seeks to intervene in the course of the development or pathogenesis of a disease. Specifically, *primary prevention* (the prepathogenic period) seeks to identify harmful habits or lack of knowledge in an otherwise healthy person and counsel that person in an effort to prevent disease; *secondary prevention* (following the clinical appearance of a disease) seeks to identify the disease at an early stage, intervene to prevent progression, and restore the patient to the predisease state of health; and *tertiary prevention* (following the appearance of a chronic disease) seeks to minimize complications and prevent further progression.
- Health protection, which is a set of activities undertaken to improve, stabilize, or otherwise manipulate the patient's environment or surroundings in order to protect and improve the patient's health.
- Environmental and engineering controls, which are controls or processes (e.g., negative pressure rooms, ventilation systems, microbial filtration devices, containers for disposal of sharp items, self-sheathing needles, etc.) that isolate or remove pathogens from the work environment or living area.
- Health education, which is any combination of learning experiences designed to facilitate voluntary adaptations of behavior that will be conducive to health.
- Personal protective equipment, which is specialized clothing or equipment that is worn by an individual for protection against pathogens or infected material.

2. General Management

- The hospital administrator and chief medical officer at each institution should develop procedures to identify and assess infectious diseases and related health risks and to implement practices and procedures to reduce the incidence and prevalence of disease and attenuating health risks.
- Practices should conform to current accepted medical standards and comply with established and published guidelines and recommendations issued by the Centers for Disease Control and Prevention (and its Advisory Committee for Immunization Practices), the Occupational Safety and Health Administration (OSHA), the National Institute for Occupational Safety and Health, and state and local departments of health.

- The health administrator and chief medical officer should establish health promotion programs encompassing disease prevention, health protection, and health education for infectious diseases. In the area of disease prevention, infectious diseases and health risks should be identified by such means as health screenings, risk assessments, physical examinations, laboratory reports, injury reports, and reviews of patient histories, food preparation and serving practices, and professional literature. In the area of health protection, procedures should be implemented to identify and reduce environmental health risks. In the area of health education, programs and courses should be offered that will help staff and inmates make informed decisions and judgments and participate actively in their own health care.
- Each institution should designate a member of the clinical health care staff (either a physician, dentist, physician assistant, nurse practitioner, or nurse) to serve as the local coordinator of infectious disease control, who should enter all cases of infectious diseases into the record and ensure that patient evaluation and follow-up are consistent with current recommendations of the Centers for Disease Control and Prevention. The coordinator of infectious disease control should also keep the state health department informed of all cases of reportable infectious diseases at the institution and should monitor prevalence and incidence information.
- Each institution should subscribe to the *Morbidity and Mortality Weekly Report*, issued by the Centers for Disease Control and Prevention. This publication contains specific guidelines and recommendations for the identification, prevention, and treatment of infectious diseases, and it should be made available in the medical library for medical staff to review.

3. Medical Testing

- Following an incident in which a staff member or an inmate may have been exposed to bloodborne pathogens, written and informed consent should be requested from the source individual to acquire and process his or her blood or other biological specimen for the purpose of determining an exposure. No inmate may be tested forcibly or involuntarily, unless there is a court order requiring such testing. Inmates may be subjected to disciplinary

action for assaultive behavior that involves exposure to bloodborne pathogens.
- Other than for incidents arising solely from possible exposure to bloodborne pathogens, as cited above, HIV testing programs are mandatory and include a yearly random sample, yearly new commitment sample, new commitment retest sample, prerelease testing (including prefurlough testing), and clinically indicated testing. Inmates must participate as ordered in all mandatory testing programs, and staff should submit incident reports on inmates who refuse to follow an order to participate in any such tests.
- Clinically indicated diagnostic procedures (other than in cases involving possible exposure to bloodborne pathogens) may be considered mandatory for inmates. An inmate who refuses routine diagnostic procedures and evaluations for infectious and communicable diseases may be subject to disciplinary action for failing to follow an order; he or she may also be subject to isolation or quarantine from the general population until assessed as not having a communicable disease or until the attending physician determines that he or she poses no health threat if returned to the general population.

If isolation is not practicable, an inmate who refuses to comply with the diagnostic process or evaluation may be evaluated or tested involuntarily. Prior to any involuntary diagnostic evaluation or procedure, the chief medical officer should advise the medical administrator at the correctional agency's headquarters, indicating the inmate's name and registration number, the specific disease for diagnosis, and the expected consequences of failing to perform the necessary diagnostic examinations. The chief medical administrator should also verify that the proposed diagnostic examinations and the reasons for them have been explained to the inmate in a format appropriate to his or her educational level and language. After documented consultation with the correctional agency's legal counsel, the medical administrator may provide the institution's chief medical officer written acknowledgment and authorization. The chief medical officer may not undertake any involuntary testing or evaluation without the specific written consent and authorization of the medical administrator.

The chief medical officer should educate and counsel the inmate prior to any involuntary evaluation or testing and carefully explain the necessity for and details of all required procedures. The diagnostic procedure or evaluation may be administered without consent, with the medical administrator's authorization, only if the counseling and education efforts are unsuccessful. Staff may use only the amount of force necessary to gain the inmate's compliance. All aspects relating to the administration of any involuntary diagnostic procedures should be documented in the inmate's medical file.

4. Infectious Disease Training

- Each institution should develop and present a comprehensive training program on infectious diseases for both staff and inmates. Because of the evolving state of medical knowledge regarding HIV in particular, the official responsible for this program should ensure that all information presented represents the most current findings and conclusions on the topic.
- A qualified health care professional should provide training to inmates during the institution's admission and orientation program. This training should include a factual presentation about infectious diseases, followed by a question-and-answer session. Additional training on infectious diseases should be provided to inmates at least once a year. Further educational programs should be developed and presented to inmates who are HIV-positive or who have contracted certain other infectious diseases, in order to assist those individuals in managing the course of their disease. A record should be maintained of all HIV-related education and training provided to each inmate.

  Infectious disease training for inmates should be appropriate in content and vocabulary to the educational level, literacy, and language of the audience, and should contain the following elements:

  —information on the epidemiology and symptoms of bloodborne diseases
  —ways in which bloodborne pathogens are transmitted
  —explanation of the exposure control plan
  —information on recognizing tasks that might result in occupational exposure
  —explanation of the use and limitations of work practice, engineering controls, and personal protective equipment
  —information on the types, selection, proper use, location, removal, handling, decontamination, and disposal of personal protective equipment

- —information on the hepatitis B vaccination, such as safety, benefits, efficacy, methods of administration, and availability
- —information on whom to contact and what to do in a medical emergency
- —information on reporting an exposure incident and on the postexposure evaluation and follow-up
- —information on warning labels, signs, and color coding, with regard to health and safety matters
- —a question-and-answer session on any aspect of the training
- Public health-related publications issued by the U.S. Department of Health and Human Services (DHHS) and its various components should be used as the primary source of educational and training materials, including *A Curriculum Guide for Public Safety and Emergency Response Workers* (DHHS–National Institute for Occupational Safety and Health publication no. 89–108) and *Core Curriculum on Tuberculosis* (DHHS–Centers for Disease Control and Prevention publication no. 199733–974, available from the U.S. Government Printing Office).

5. Standards and Compliance

- Infectious waste should be handled, controlled, and disposed of in accordance with standards issued by OSHA, as published in the *Code of Federal Regulations* (29 CFR #1910.1030). Plans and procedures regarding bloodborne pathogens should comply with OSHA standards set forth in the *Code of Federal Regulations* (29 CFR #1910.1030) and any applicable state regulations or statutes.
- All sexually transmitted diseases should be treated in accordance with treatment guidelines issued by the Centers for Disease Control and Prevention, as published in the *Morbidity and Mortality Weekly Report*, volume 38, number S–8, September 1, 1989, or in accordance with any successor guidelines that may be issued.
- Training records relating to OSHA standards should be retained for at least three years and may be made available upon request to the director of the National Institute for Occupational Safety and Health and the assistant secretary of labor for occupational safety and health.

6. Medical Isolation and Quarantining

- The chief medical officer, in consultation with the hospital administrator, should ensure that inmates with infectious diseases

that are transmitted through casual contact (including tuberculosis, chickenpox, measles, etc.) are isolated from the general inmate population until such time as they are assessed or evaluated by a health care provider.
- Inmates should remain in medical isolation, as ordered by the chief medical officer, until they pose no threat to others, unless their activities, housing, and duty assignments can be limited, or unless environmental and engineering controls or personal protective equipment can eliminate risk of transmitting the disease. It is the responsibility of the chief medical officer to determine the appropriateness for duties and housing of inmates demonstrating infectious diseases that would be transmitted through casual contact.
- Limitations may be placed on duties and housing assignments only for inmates whose infectious diseases can be transmitted despite the use of environmental and engineering controls or personal protective equipment, or if precautionary measures cannot be implemented or are not available to prevent the transmission of the specific disease. The warden, in consultation with the chief medical officer, may exclude inmates from work assignments, on a case-by-case basis, based upon the type of institution and the need to maintain institutional operations that are safe and orderly.
- Special housing for inmates who are HIV-positive should not be used, except in cases where an HIV-positive inmate poses a danger to others (as outlined in "HIV-Positive Inmates Who Pose a Danger to Others," below).

7. Confidentiality of Information

- Information related to any inmate's medical condition—including information about infectious diseases—should be regarded as confidential. Access to such information should be restricted to the medical staff or other staff having a legitimate need to know. Any release of such information should be made only in accordance with applicable Freedom of Information and privacy statutes.
- Medical information concerning the chronic infectious diseases of individual inmates should be limited to members of the institution medical staff, the institution psychologist, and—in order to address issues regarding prerelease and postrelease manage-

ment—the warden, case manager, probation or parole officer, and halfway house director.
- When an employee has been exposed to possible bloodborne pathogens from an inmate, information about the inmate's medical condition may be divulged to that employee's own physician if he or she can show a compelling need to have that information in order to treat the employee, and provided release of the information is authorized under applicable Freedom of Information and privacy statutes. In such cases, the chief medical officer should seek consent from the inmate who is the source individual and advise the employee's physician of the necessity for maintaining all medical information regarding the source individual in strictest confidence. When the exposed individual is an inmate, he or she will be treated by institution staff, who will have access to medical information on the source individual. Medical information on the source individual should not be divulged to the inmate who has been exposed.

8. Tuberculosis Control

Because of the close confinement of individuals in correctional facilities, and the fact that tuberculosis can be transmitted easily when an infected individual coughs, speaks, or spits, inmates and staff may be susceptible to the disease. Prevention and control procedures need to be in place to protect staff and inmates alike. Institutions should comply with recommendations issued by the Centers for Disease Control and Prevention in two publications: *Control of Tuberculosis in Correctional Facilities: A Guide for Health Care Workers* and *Guidelines for Preventing the Transmission of Tuberculosis in Health Care Facilities*.
- Tuberculin skin testing (Mantoux method) should be provided to all institution employees and inmates yearly, and this testing should be consistent with guidelines established by the U.S. Department of Health and Human Services. The health care provider evaluating the skin test should document the results in the employee's or the inmate's medical record.
- Any employee who tests positive should be referred to his or her private health care provider; further, the employee should sign a release of information form permitting his or her health care provider to notify the institution's chief medical officer of any treatments and recommendations for care. The employee's duty

assignments may be adjusted in order to minimize potential risks of transmission of tuberculosis or threat to the health and well-being of other employees or inmates.
- Positive skin tests for either employees or inmates should be reported to the appropriate state health department, in accordance with applicable state regulations, and examining medical personnel should document all known recent exposures. The chief medical officer should evaluate and counsel all employees and inmates with positive skin tests.
- If an inmate is clinically suspected of having tuberculosis, as diagnosed by a primary health care worker, he or she is to be removed from the general population immediately. Although a person presenting clinical indications may not have tuberculosis, he or she should be treated as if infected until it is confirmed that the disease is not present or until treatment renders the disease noninfective. If the institution has a negative pressure isolation room, then the inmate should be placed there until arrangements can be made to transport and admit the inmate to a local hospital with facilities to treat tuberculosis cases. If the institution does not have this room, then the inmate should be placed in a room with outside ventilation and outside exhaust until arrangements for transportation to a local hospital can be made.
- When occupied by a suspected tuberculosis case, an isolation room should have a warning sign posted outside. Any staff member entering the room of the suspected tuberculosis patient should wear a high efficiency particulate air (HEPA) respirator, approved by the National Institute of Occupational Safety and Health.
- Special precautions should be taken when transporting an inmate suspected of having tuberculosis to the hospital. All staff coming into contact with the inmate, including escort officers, medical staff, and receiving and discharge personnel should wear HEPA respirators. HEPA respirators are not safe for a person potentially infected with tuberculosis, so the inmate should wear a standard surgical-type mask. No other inmates should be transported on the same trip. Contract guard staff who may be hired to provide security for the inmate at the hospital should be issued HEPA respirators.
- If an inmate with clinically suspected tuberculosis is scheduled to appear in court or at a hearing of the parole board, the Immigration and Naturalization Service, or other body, the warden should notify the appropriate hearing authority that the inmate is undergoing treatment for clinically suspected tuberculosis and cannot

be moved until a determination of infectivity has been made. If possible, a tentative treatment timetable and date of inmate availability should be given to the hearing authority.
- Any case of an inmate being identified with clinically suspected tuberculosis should be reported to the correctional agency's headquarters. The report should include the diagnostic rationale for identifying the person as a suspected case and note steps taken to ensure respiratory isolation. If a suspected case is found to have active tuberculosis, this information should be entered into the automated data system. A positive skin test must not be entered into the database as a positive tuberculosis case.

9. HIV and Hepatitis B Virus Testing and Treatment

- During routine intake screening, all new commitments should be interviewed to identify those who may be infected with human immunodeficiency virus (HIV) or hepatitis B virus (HBV). The interviews should be conducted by medical staff and should address specific symptoms such as thrush, nausea, fevers, chills, night sweats, cough, unexplained weight loss, lymphadenopathy, and diarrhea. The interviews also should include questions regarding high-risk behaviors, sexual activity, intravenous drug use, and blood transfusions. Medical staff may request inmates identified as being at high risk for HIV or HBV to submit to tests for those conditions. Those who refuse may be subject to disciplinary action for failing to follow an order. It is the responsibility of medical staff to determine if HIV or HBV testing is clinically indicated.
- During a given period once a year, all newly incarcerated inmates in institutions throughout the correctional agency should be tested. Those who test negative should be entered into a retest group and tested twice a year thereafter. Testing for the new commitment group and the retest group is mandatory, and failure to submit to testing may result in disciplinary action for failure to follow an order.
- A random sample for HIV/HBV of all inmates in the system should be conducted once a year. Inmates tested in this random sample should not be scheduled for follow-up routine testing. The chief of research at the correctional agency's headquarters should determine the random method to be employed in selecting the sample.

- After consultation with a member of the institution's medical staff, an inmate may request an HIV/HBV antibody test. Ordinarily, inmates may not ask to be tested more than once a year.
- An institution physician may order an HIV/HBV antibody test if an inmate has chronic illnesses or symptoms suggestive of an HIV or HBV infection. Inmates who are pregnant, inmates receiving live vaccines, or inmates being admitted to community hospitals should be tested. Inmates demonstrating sexual behavior that is promiscuous, assaultive, or predatory should be tested. Testing is also indicated for opportunistic infections and certain cancers, as well as some patients with a positive purified protein derivative (PPD) test, Venereal Disease Research Laboratory (VDRL) test, or a history of any sexually transmitted disease.
- Inmates being considered for full-term release, parole, good conduct time release, furlough, or placement in a community-based program such as a halfway house should be tested for HIV/HBV. An inmate who has been tested within one year of this consideration ordinarily would not be required to submit to a repeat test. An inmate who refuses to be tested may be subject to disciplinary action and ordinarily should be denied participation in any community-based programs. Emergency furloughs, however, may be considered if current HIV and HBV antibody test results are not available. A seropositive test result may not be considered as the sole grounds for denying participation in a community-based program or activity.

No later than thirty days before the scheduled release for parole or placement in a community-based program of an inmate with positive HIV or HBV status, the warden should send a letter to the appropriate supervising official in the location where the inmate is to be released, advising that official of the inmate's positive HIV or HBV status. A copy of this letter should also be forwarded to the community programs manager at the correctional agency's headquarters, who should notify the director of the halfway house (if applicable). Similarly, prior to the participation of an HIV- or HBV-infected inmate in any community activity (including a furlough), the warden should notify the supervising official in the area to be visited of the inmate's status, and the hospital administrator should notify the state health department. Prior to release of an HIV- or HVB-positive inmate to an Immigration and Naturalization Service detainer, the warden should notify the appropriate official at that agency. In all instances of notification, precautions should be taken to ensure that only authorized persons with a legitimate need to know are allowed access to the information. Prior to release on parole, completion of sentence, placement in a community-based program, or participation in an unescorted community activity, an HIV-positive or HBV-infected inmate should be strongly encouraged by the hospital administrator to notify his or her spouse or other identified significant others with whom the inmate might have contact, risking possible transmission of the virus.

- Initial HIV and HBV blood tests should be conducted using the Enzyme-Linked Immunosorbent Assay (ELISA). Positive ELISAs should be repeated on the initial sample. If the second ELISA is positive, a confirmatory Western Blot should be performed on the same sample. If the Western Blot is indeterminate or contradicts the ELISA result, a second sample should be drawn within thirty days. Only a positive ELISA result, confirmed by a Western Blot, should be entered into an inmate's medical file. The institution physician should classify all inmates with a confirmed positive HIV or HBV antibody test using the appropriate Centers for Disease Control and Prevention classification and following guidelines issued by the Centers for Disease Control and Prevention.
- Testing information should be entered into an automated database, to ensure accurate tracking and facilitate statistical analysis. Access to medical information on specific individuals is restricted to medical staff at the institution and the correctional agency's headquarters. Reporting of seropositive test results should serve as the basis for formulating policies for the care and treatment of AIDS, HIV, and HBV cases; training and management policies; and fiscal plans.
- A seropositive test result should not, by itself, constitute grounds for disciplinary action. Disciplinary action may be considered when it is suspected that the inmate has engaged in a secondary action that may have risked transmission of the virus, such as sharing razor blades.
- HIV and HBV testing should be administered in conjunction with an education and counseling program. Inmates receiving the HIV or HBV antibody test should receive pretest and posttest counseling, regardless of test results. Counseling should address the limitations of the test (including false positives, false negatives, and the possible need for additional testing) and also cover the complications and consequences of a negative or positive test

result. Medical staff should counsel inmates testing positive on how to maintain their health and avoid transmitting the virus to others. Pregnant inmates who test positive should also be advised as to the likelihood of the virus being transmitted to the fetus. Inmates testing positive should also be referred to the Psychology Department for follow-up counseling.
- Clinical evaluation and review of each HIV-positive inmate should be conducted by hospital staff at least once a year. The prevention and treatment of HIV and HBV infections serve to restrict the spread of those diseases and can maintain the quality of life for those suffering from them; therefore, hospital staff should monitor HIV- and HBV-positive inmates closely and pay particularly close attention to tests that can determine the status of the inmate's immune system. When clinically indicated, pharmaceuticals approved by the U.S. Food and Drug Administration for use in the treatment of AIDS, HIV-positive, and HBV-positive cases should be offered at the institution.
- When an HIV- or HBV-positive inmate is transferred to another facility, medical staff should note under the special instructions section of the transfer order that "CDC Universal Precautions are to be observed when transporting this inmate," and the warden of the sending facility should notify the warden of the receiving facility of the inmate's medical status.
- Staff may evaluate inmates who are seriously ill as a result of their HIV or HBV infection for possible resentencing, under provisions of the correctional agency's policy on compassionate release, if authorized by state law.

10. HIV-Positive Inmates Who Pose a Danger to Others

- Inmates who have tested HIV positive may be placed in controlled or special housing status if there is reliable evidence indicating that they are going to engage in conduct, or are continuing to engage in conduct, that poses a specific health risk to another person. Such action should be considered only when an HIV-positive inmate poses a threat to others and must not be imposed as a standard method for handling inmates who test positive.
- HIV is not easily transmitted. It is not spread through casual contact, such as ordinary contact in family settings, prisons, schools, or other groups living or working together. Nor is it spread through common use of drinking cups, eating utensils, or toilet seats. In fact, the available evidence indicates that the virus

is transmitted primarily through two very specific routes: blood-to-blood contact (through such practices as sharing needles) and sexual fluids to blood contact (occurring in sexual practices involving the passage of fluids). As part of its infectious disease control program, the institution should have education and counseling components in place to advise HIV-positive inmates on avoiding high-risk behavior that would endanger others.

- The warden may submit a recommendation to the appropriate executive-level official at the correctional agency's headquarters (a deputy commissioner or deputy director) that an HIV-positive inmate be considered for placement in special housing status, if that inmate engages in or indicates a disposition to engage in high-risk behavior, despite counseling and education on avoiding such behavior. Evidence that an HIV-positive inmate is a threat to others in this way would include statements made by the inmate, repeated instances of misconduct, or other actions suggesting that the inmate may share needles with other persons, engage in predatory or promiscuous sexual behavior, or exhibit assaultive behavior where body fluids may be transmitted to another person.
- The warden's recommendation should be in writing, should cite the basis for the referral, and should be accompanied by a current progress report on the inmate, an up-to-date mental health report, and a comprehensive medical report (which should include confirmation of the inmate's HIV-positive status).
- Based on the perceived health risk posed by the inmate's threatened or actual behavior, the warden may, with the approval of the deputy director or deputy commissioner, place the inmate in administrative segregation, a secure hospital room, or other form of special housing for a period not to exceed twenty working days, pending the inmate's appearance before the special housing hearing administrator. Reasons for this placement, and verification of the authorization from the deputy director or deputy commissioner, should be placed in the inmate's central file. The inmate should be seen daily by case management and medical staff while on this temporary special housing status, and a mental health assessment should be prepared during this period.

If the referral is prompted by an incident report on the inmate, he or she may be placed in administrative segregation in accordance with the institution's disciplinary policy (see Chapter 2, Part VIII). If found to have committed an infraction, the inmate may be placed in disciplinary detention for a prescribed period and thereafter may be placed in postdisciplinary detention for up to ninety days.

- In all other regards, including hearings, the decision-making process, reviews, the inmate's right to appeal, release from special housing, etc., the process of referring an inmate for possible placement in special housing should adhere to institution policies governing all special housing placements (see Chapter 2, Part VII). Conditions within special housing for HIV-positive inmates placed in special housing status for posing a threat to others should also conform to the policy on special housing units. In particular, inmates under special housing status should be considered for the same activities, programs, and services as inmates in the general population, to the extent consistent with the institution's available resources and security needs.

II. SUICIDE PREVENTION PROGRAM

Overview

Although it would be impossible to eliminate all suicides from any population, including a prison population, each institution should implement a suicide prevention program in order to carry out its mission of monitoring and protecting the health and welfare of individuals in its custody. A suicide prevention program should include a staff training component, inmate screening, and appropriate preventive supervision, counseling, and other treatment for inmates clinically found likely to be suicidal.

1. Program Coordinator

- The warden should designate a full-time staff member to serve as program coordinator for the institution's suicide prevention program. The program coordinator's responsibilities would include managing the treatment of suicidal inmates and for ensuring that the institution's suicide prevention program conforms to policy guidelines for staff training, identification of potential suicide cases, referral of suicide cases, and assessment or intervention. Ordinarily, the institution's chief of mental health services should serve as program coordinator, but he or she may delegate program responsibility to another mental health professional who is responsible for providing clinical management of cases where suicide risk assessment and intervention are indicated.

2. Staff Training

- The initial period of incarceration is often a critical time for detecting potential suicide cases, but serious suicide crises may arise at any time. Signs of potential suicidal behavior often are first identified by line staff, based on their frequent interactions with inmates. Therefore, it is imperative that staff receive training, ordinarily from mental health personnel, on how to recognize signs of a potential suicide case, on the appropriate referral process, and on suicide prevention techniques.
- Suicide prevention training should be included in the basic training curriculum at the staff training academy, during institution familiarization training, and in annual refresher training at all institutions. Training should focus on identifying suicide risk factors, typical inmate profiles of completed suicides, recognition of potentially suicidal behavior, and methods of referral. In addition, supplemental specialty training in suicide prevention should be provided to physician assistants and nurse practitioners, shift commanders, and counselors, as well as to all staff who work with inmates in special housing units.
- Wardens should periodically include discussions of suicide prevention at department head meetings, staff recalls, etc., to remind staff of the need to observe inmates constantly for signs of suicidal behavior.

3. Identification and Referral of Potential Suicide Cases

- All newly admitted inmates should be screened by a physician assistant, normally within twenty-four hours of admission to the institution, for signs of potential for suicide. During intake screening, if a physician assistant, nurse practitioner, or other staff member notes signs that an inmate is potentially suicidal, he or she should refer the inmate immediately to the program coordinator for further evaluation.
- Except for emergency cases that are referred to the program coordinator immediately (as noted above), mental health staff

should conduct more comprehensive assessments of inmates within fourteen days of their admission (if they are newly incarcerated) or within thirty days of their admission (if they are transfers from other facilities).
- Staff members must never take any inmate's suicide threats or attempts lightly, or take lightly any information or suggestions from other inmates that an inmate may be potentially suicidal. During regular working hours, staff members should advise the program coordinator immediately of any inmate who exhibits behavior indicative of suicide potential. In emergency situations or during nonregular working hours, staff members should advise the shift commander, who should place the inmate on a formal suicide watch pending evaluation by the program coordinator.

4. Assessment and Intervention

- During regular working hours, inmates referred for assessment of suicide potential should be seen on a priority basis. During nonregular working hours, the program coordinator should consult with institution staff and may choose to see the inmate immediately or have the inmate placed on a suicide watch. In either case, the inmate should receive an individual assessment within twenty-four hours. Once an inmate has been placed on a suicide watch, it may not be terminated under any circumstances without an in-person evaluation being performed by the program coordinator (or designee) or other mental health professional. The program coordinator should notify the mental health administrator at the correctional agency's headquarters whenever a suicide watch is initiated or terminated, and when a formal suicide watch exceeds seventy-two hours.
- If, upon evaluating an inmate referred for assessment, the program coordinator determines that the inmate does not appear imminently suicidal, the inmate may be removed from suicide watch. The program coordinator should document in writing the basis for this conclusion and any treatment recommendations. Copies of this documentation may be placed in the inmate's medical file, psychology file, and central file. A suicide risk assessment form may be developed to facilitate such documentation.
- If the program coordinator determines that the inmate demonstrates an imminent potential for suicide, the inmate should remain on suicide watch in the institution's designated suicide prevention

room (see below). The program coordinator should document his or her actions or findings (on a suicide risk assessment form, if applicable), and copies of this assessment should be placed in the inmate's medical file, psychological file, and central file.

- Inmates on suicide watch should be placed in the institution's designated suicide prevention room—a separation cell, preferably located in the institution hospital. Another site may serve as the suicide prevention room if it is not feasible to place it in the institution hospital, but under no circumstances should a suicide prevention room be located in administrative segregation or in a general housing area. Although the suicide prevention room should be located within the institution hospital, if possible, an inmate who is confined there should not be admitted as an inpatient unless there are medical indications that would necessitate immediate hospitalization.
- The suicide prevention room should be set up in such a way as to permit the constant observation, protection, and control of the inmate. There must be an unobstructed view of and easy access to the inmate at all times. While it is impossible to make any area completely suicide-proof, every effort should be made to remove or modify fixtures, articles, or architectural features that could be used for self-injury. The room should contain only limited and secured furnishings.
- While an inmate is on a formal suicide watch, the conditions of confinement should be the least restrictive available that would ensure control and safety. Security staff, in consultation with the program coordinator or designee, should be responsible for the inmate's daily custodial care and routine activities. Unit staff (including members of the inmate's unit team) should also be available to respond to the needs of the inmate while he or she is on suicide watch.
- The program coordinator should have responsibility for determining specific conditions of the suicide watch. He or she should specify the type of personal property, bedding, clothing, magazines, or smoking materials that should be allowed. Deviations from the prescribed living conditions should be made only with the program coordinator's concurrence.
- The program coordinator should interview or monitor each inmate on suicide watch at least once a day and record clinical notes following each visit. Unit staff should also have frequent contact with the inmate. Individuals assigned to suicide watch should maintain constant observation of the suicidal inmate and be able

to maintain verbal communication with the inmate at any time. The suicide watch may be conducted either by institution staff or, when authorized by the warden, trained inmate "companions" chosen by the program coordinator (see below).
- The observer conducting the suicide watch should not be in the same room with the inmate, unless he or she must enter the suicide prevention room in an emergency or to provide specific services. Rather, the individual on suicide watch should observe the suicidal inmate through a window, and there should be a locked door between the inmate and the observer.
- The person performing the suicide watch should have a means to summon help immediately if the inmate displays any suicidal or other unusual behavior. Restraints may be applied if necessary to gain greater control over an inmate who is under a suicide watch, but the use of restraints should be clearly documented and supported.
- Staff assigned to suicide watch must have received specific training and must review and sign the post orders before starting the watch. The program coordinator should review the post orders annually and revise them as appropriate.
- At the warden's discretion, an institution may utilize inmates as companions to help monitor suicidal inmates. The program coordinator would be responsible for selecting inmate companions, providing them with semiannual training in program procedures and purposes, assigning inmate companions to specific assignments, and removing them as necessary. Actual authorization for the use of inmate companions should be made by the warden, on a case-by-case basis.

A sufficient number of companions should be trained to provide round-the-clock observation of potentially suicidal inmates, and alternate candidates should always be available. Lists of inmates authorized to serve as companions should be available to the security staff during nonregular working hours.

Companions should be selected based upon their ability to perform the specific task and also for their reputation within the institution. They should be mature and reliable individuals who have credibility with both staff and inmates, can perform their duties with a minimal need for supervision, and can be relied upon to protect the suicidal inmate's privacy from other inmates.

Except under unusual circumstances, inmate companions should not work more than one four-hour shift in any twenty-four-hour period. Inmate companions should sign an agreement of understanding and expectations and may receive performance pay for time spent monitoring a potentially suicidal inmate. The program coordinator should maintain a file on each inmate companion, containing signed agreements of understanding and expectations, documentation of training sessions, and verification of pay for those who performed watches.

Supervision of inmate companions performing suicide watches should be carried out by staff in the immediate area and should include checks at least once every hour. Staff members should initial the suicide watch logs (see below) upon conducting checks. In no case should an inmate companion be assigned to a watch without adequate provisions for staff supervision or without the ability to obtain rapid staff assistance.

During each suicide watch, a chronological log should be maintained by the observer to record the inmate's behavior and other relevant information, at fifteen-minute intervals. Staff visits to the inmate should also be recorded in the log. The program coordinator should review and initial the log daily.

- Only the program coordinator has the authority to remove an inmate from suicide watch. Termination of the watch should be documented, with copies of the documentation sent to the warden and placed in the inmate's medical file, psychological file, and central file. Documentation should include a clear description of the resolution of the crisis and guidelines for follow-up care.

5. Custodial Issues

- Inmates in administrative segregation or disciplinary detention status often may be at higher risk for suicidal behavior. The program coordinator should make weekly rounds of the special housing units and consult with staff in those areas concerning inmates needing special attention. Staff members assigned to the special housing units should be alert to signs that an inmate may be potentially suicidal, and the program coordinator should arrange for them to receive special suicide prevention training beyond that provided to other staff. Indications of possible suicidal behavior should be reported immediately to the program coordinator.

- The program coordinator should arrange to remove any inmate from special housing or disciplinary detention status who exhibits potential for suicide and to place that inmate on a suicide watch. Once the suicide crisis is over, the inmate should be returned to the special housing unit to satisfy any sanction that was imposed, unless the segregation review official finds that completion of the sanction is no longer necessary or advisable.
- Engaging in an act of self-injury does not necessary indicate suicidal intent on the part of an inmate, nor does it absolve an inmate of responsibility for the consequences of his or her behavior. Thus, an incident report may be prepared by a staff member who discovers any self-injury or attempted self-injury, citing the inmate for engaging in the prohibited act of self-mutilation, for engaging in conduct that disrupts or interferes with the orderly running of the institution, or for other violations.

If an incident report is written, the program coordinator should review the case and render an opinion as to the inmate's responsibility for his or her behavior at the time of the act and address the issue of the inmate's competency. Based upon this opinion, appropriate administrative or disciplinary action may be taken.

6. Transfer of Inmates to Other Institutions

- The program coordinator may refer a suicidal inmate to a medical center operated by the correctional agency or a community health care facility at any time. In addition, the program coordinator is required to consider referral for any inmate who has been on a continual suicide watch for seventy-two hours. As part of the referral consideration process, the program coordinator may consult with other mental health professionals, staff at the medical center or community health care facility, or the mental health administrator at the correctional agency's headquarters. Referral considerations and all actions taken regarding possible referrals should be documented in the inmate's medical file, psychological file, and central file.
- No inmate who is determined to be imminently suicidal may be transferred to another institution, except on an emergency basis to a medical center. Inmates who are not imminently suicidal, but have a history of potentially suicidal behavior, may be subject to

routine transfers. In such cases, the program coordinator at the sending institution should notify the program coordinators at any holdover institutions and at the receiving institution concerning the inmate's history.

7. Analysis of Suicides

- If an inmate suicide does occur, the program coordinator should notify the mental health administrator at the correctional agency's headquarters immediately. The mental health administrator should arrange for a psychological reconstruction of the suicide to be conducted by a psychologist from another institution. The scene of the suicide should be protected in the same manner as any crime scene in which a death has occurred. All measures necessary to preserve and document the evidence needed to support subsequent investigations should be maintained or otherwise recorded.
- The psychologist conducting the psychological reconstruction of the suicide should submit a written report of his or her conclusions to the mental health administrator at the correctional agency's headquarters.

III. FOOD FROM OUTSIDE SOURCES

Overview

Institutions are responsible for ensuring that regular meals served to inmates on the main line, as well as special diets served to inmates for medical reasons or to conform to religious requirements, are sanitary and meet nutritional guidelines. In addition, under certain conditions food may be brought into the institution from outside sources; the institution is responsible for ensuring the quality of that food as well. Food brought into the visiting area in accordance with visiting procedures and food items sold in the institution commissary are exempt from such requirements.

1. Special Meals Prepared by the Food Service Department

- With thirty days notification, the institution's Food Service Department can provide banquet, holiday, and other special meals, as well as special meals requested by an inmate organization (see

Medical and Psychological Issues

Chapter 4, Part IV). If the warden approves a request for a banquet, holiday, or other special meal, the institution's chief of food services should take full responsibility for the planning and serving of the meal.
- Meals requested by a group for a special event may be planned on the master cycle menu, either exactly as requested or with minor modifications.

2. Inspection of Outside Meals

- Meals brought in by any outside or inside organization should be inspected carefully both by security staff and the sponsoring department. A method of inspection to search food items should be approved by the chief of food services (for proper preparation and storage) and the chief of security (for contraband), before the food items can be handled by inmates. Particular attention should be given to bulk purchases, such as pizzas, sandwiches, and cakes.
- Careful attention should be given to the handling of food items to ensure that inspection methods do not alter the food's quality, sanitation level, or content, and that appearance is altered as little as possible.

IV. INMATE HUNGER STRIKES

Overview

At any institution, there may be instances where an inmate will threaten or actually initiate a hunger strike, for any number of reasons. Such incidents should be taken very seriously by staff. It is the responsibility of the institution to monitor the health and welfare of inmates in its custody and to pursue procedures designed to preserve life; therefore, the institution should assume medical and administrative management of all inmates who engage in hunger strikes. Inmates with metabolic disorders or certain other illnesses who deviate from normal eating habits or intake of fluids could experience an immediate and significant hazard to their health and well-being. For other inmates, a deviation from regular eating and drinking habits could indicate a mental disorder. In any case, long-term deprivation of food and shorter-term deprivation of fluids can cause serious and possible irreversible physiological changes and can even lead to death.

1. Identification and Referral

- An inmate may be considered to be on a hunger strike when he or she communicates that fact to staff or is observed by the staff to have refrained from eating for at least seventy-two hours. At times, however, an allegation of a hunger strike may be made that is not reflected in any overt action and is merely a bid to gain attention.
- An inmate observed to be on a hunger strike should be referred to medical or mental health staff for evaluation and possible treatment. Although referral ordinarily would not occur until after the inmate has refrained from eating for at least seventy-two hours, there may be occasions where it may be prudent to initiate management and medical intervention before seventy-two hours have elapsed.
- Medical staff should place the inmate in a medically appropriate, locked, single-cell observation room for close monitoring. The inmate should not have contact with other inmates while in this room, and all monitoring should be conducted by staff members. Inmate companions (such as those used to monitor inmates on suicide watch) should *never* be used to monitor an inmate on a hunger strike.
- An inmate in special housing while on administrative segregation or disciplinary detention status may remain there for monitoring while on a hunger strike, unless the physician determines that he or she must be moved to other quarters for medical reasons.

2. Initial Medical Evaluation and Management

- Medical staff immediately should examine all inmates referred for being on hunger strikes, by conducting a general medical evaluation; measuring and recording the inmate's height and weight; taking and recording vital signs; performing a urinalysis; conducting a mental health evaluation; taking radiographs, as clinically indicated; and conducting laboratory studies, as clinically indicated. If an inmate refuses to submit to medical evaluations, staff should request the inmate to sign a refusal of treatment form. If the inmate refuses to sign the form, staff members should document the refusal.
- Medical staff members should measure and record weight and vital signs at least once every twenty-four hours while the inmate is on the hunger strike. Other examination procedures noted

above should be repeated, modified, expanded, or discontinued, as medically indicated. All medical procedures conducted upon an inmate on a hunger strike must be documented in the inmate's medical file.
- An inmate on a hunger strike may be transferred to one of the correctional agency's medical centers, to another institution in the system, or to a community health care facility, if deemed necessary for medical reasons.

3. Intake and Output of Food and Fluids

- The staff must deliver three meals a day to the inmate's room. Additional meals may be sent at other intervals, as authorized by the physician. The meals must be physically placed in the inmate's room; it is not sufficient for staff to offer meals verbally.
- The staff should provide the inmate with an adequate supply of drinking water. Other beverages should be offered, but the inmate's acceptance of beverages other than drinking water should be documented and medical staff should be advised.
- Staff should remove any commissary food items or other private food supplies from the possession of any inmate who is on a hunger strike. An inmate on a hunger strike should not be permitted to purchase food items from the commissary, but may purchase other items.
- All food and fluid intake and output should be monitored as recorded, if deemed necessary by the physician. Therefore, a dry cell (i.e., a cell without plumbing) must be available for housing an inmate on a hunger strike.

4. Forced Feeding

- If a physician determines, based on the inmate's inadequate intake or abnormal output, that the inmate's life or health would be threatened if treatment were not initiated immediately, he or she may consider forced medical treatment of the inmate (i.e., forced feeding). While the decision to force treatment upon the inmate is a medical one, it has important legal implications. The physician must be convinced with reasonable medical certainty that the inmate's life is in imminent danger or that there is a strong threat of permanent damage to the inmate's health.
- Unless there already is some arrangement with the local courts that would authorize forced medical treatment, the institution

should refer the matter to the correctional agency's legal counsel when it appears to medical staff that the inmate's condition is deteriorating to the point where intervention may be required.
- Before administering medical treatment against an inmate's will, staff members must make reasonable efforts to convince the inmate to accept treatment voluntarily. Staff members also should explain to the inmate the medical risks he or she would incur by not accepting treatment voluntarily and should document their efforts to persuade the inmate in the inmate's medical file.
- If reasonable efforts fail to persuade the inmate to accept treatment voluntarily, or if an emergency exists making such efforts impossible and it becomes medically necessary to treat a life-threatening or health-threatening situation immediately, the physician may order that treatment be administered without the consent of the inmate. Treatment efforts should be documented in the inmate's medical file, and written reports on such treatments should be submitted to the deputy director or deputy commissioner at the correctional agency's headquarters who is responsible for supervising medical operations. The warden should notify the sentencing judge and the prosecuting attorney's office of any forced treatment and provide an explanation of why the forced treatment was deemed necessary. The outcome of the hunger strike and the forced treatment should also be reported.
- Only the physician is permitted to order forced medical treatment. Forced medical treatment normally should involve use of a nasogastric tube for feeding. If unsuccessful or medically inappropriate, then intravenous fluids and intravenous hyperalimentations may be necessary. As a last resort, gastrostomy and tube feeding through the stomach may be required, but such treatment should be undertaken only after seeking review from the appropriate court.
- Staff members should continue clinical and laboratory monitoring as necessary, until the inmate's life or permanent health are no longer threatened. Treatment should continue until adequate oral intake of food and fluid is achieved. The staff should continue medical and mental health follow-ups as long as necessary.

5. Authority

- Only the physician may order that an inmate may be released from hunger strike evaluation and treatment. Any such orders

must be documented in the inmate's medical file. The exercise of sound medical judgment by the physician should be the prevailing consideration at all times in dealing with an inmate on a hunger strike, and none of the policies or procedures outlined above should limit or override the medical judgment of the physician.
- Each case must be evaluated on its own merits and individual circumstances by the physician. Treatment should be given in accordance with accepted medical practice.

V. SEXUAL ASSAULT PREVENTION AND INTERVENTION

Overview

Sexual assault is forceful or intimidating behavior by one or more inmates—including pressure, threats, or other actions or communications—that causes or is meant to cause another inmate to engage unwillingly in a partial or complete sexual act. Although only a small percentage of inmates are sexually aggressive and seek to dominate other inmates through violent sexual behavior, forceful and pressured sexual interactions are among the most serious threats to inmate safety and institutional order. Victims may suffer physical and psychological harm and could be infected with serious and even life-threatening diseases. Each institution should have a program in place to prevent sexually assaultive behavior, provide assistance to victims of sexual assault, investigate allegations of sexual assault, and to discipline, control, and prosecute assailants as quickly as possible.

1. Program Coordinator

- The warden should appoint a staff member—preferably a staff psychologist—to serve as the sexual assault prevention and intervention program coordinator. A nonpsychologist may be appointed as alternate coordinator, provided that person is a supervisor; is trained in sexual assault crisis issues; and has the knowledge, skills, and abilities for program implementation and evaluation.
- The program coordinator is responsible for providing education and training to staff on sexual assault prevention and intervention; for assessing support and treatment services provided to inmates who have been sexually assaulted, and ensuring that

those services are adequate; and for overseeing all aspects of the program.

2. Prevention

- The prevention component of the program should include inmate screening and identification, inmate education, and staff training.
- All newly admitted inmates should be screened by medical staff, mental health staff, and unit staff—ordinarily within twenty-four hours of admission. One purpose of this screening is to identify inmates who have been victims of sexual assault at any time before their current confinement. Such inmates should be referred to the program coordinator for assessment and possible treatment.
- Staff should attempt to identify sexually assaultive inmates. During the intake screening and initial classification process, unit staff members should review the presentence report and central file carefully for any documentation that an inmate has a history of sexually aggressive behavior. Inmates with a history of sexually abusive or assaultive behavior should be referred to mental health staff for assessment.
- The psychology administrator at the correctional agency's headquarters should provide technical assistance and training information to institution program coordinators for a course of inmate education, including literature for distribution to inmates on how to prevent and avoid sexual assault. As part of the HIV-AIDS discussion in the inmate admission and orientation program, medical staff should include a brief and candid presentation about the sexual assault prevention and intervention program, including information on how inmates can protect themselves from becoming victims while incarcerated. This presentation should also include information on services and programs (including counseling and sex offender treatment) for sexually assaultive or aggressive inmates.
- All staff members in an institution must be alert to signs of potential situations in which sexual assaults might occur. Unit staff, security staff, and mental health staff have particularly instrumental roles in preventing sexual assaults. Staff training in sexual assault prevention and intervention, therefore, is crucial.

The program coordinator should ensure that all staff members are trained to recognize the signs of sexual assault, understand the identification and referral process when a sexual assault is alleged to have occurred, and have a basic understanding of sexual assault prevention and response techniques. For new employees, a presentation on sexual assault prevention and intervention should be part of the initial training and should include information on the correctional agency's policies on sexual assault and staff responsibilities in preventing and reporting sexual assaults. Additional information about the program should be included as part of annual refresher training. More intensive training should be provided to unit staff, including information on the identification and monitoring of high-risk areas in the unit, because unit staff are involved in the daily monitoring of inmate activities and interact with inmates during the evening and night shifts when assaults are most likely to occur.

3. Intervention

- Staff sensitivity toward inmates who are victims of sexual assaults is critical. Staff should take seriously all statements or indications from inmates that they have been victims of sexual assaults and respond supportively and nonjudgmentally. Any inmate who indicates that he or she has been sexually assaulted should be offered immediate protection and be referred for a medical examination, as well as a clinical assessment of the potential for suicide or other related problematic behavior.
- When an inmate or inmates are the alleged perpetrators, the chief of security should notify the appropriate investigative agency. When a staff member or members in the institution are the alleged perpetrators, the warden should refer the case to the correctional agency's Office of Internal Affairs, which should coordinate further law enforcement referrals. For other circumstances (such as an assault while an inmate is out of the institution on a writ or in a halfway house), appropriate law enforcement officials should be contacted.
- When an inmate has been sexually assaulted, or claims to have been sexually assaulted, staff members should notify the shift commander immediately. The shift commander should notify all other appropriate staff, including medical staff, the chaplain, the chief of mental health services, the deputy warden with responsi-

bility over security issues, and the program coordinator. During evening and night shifts, when the potential for assaults is greater, the shift commander should also notify the duty officer.
- The victim should be advised of the importance of getting help to deal with the assault and that trained staff members are available to assist. If the victim is fearful of being labeled an informer, he or she should be advised that it is unnecessary to reveal the identity of the assailant in order to receive assistance. At a minimum, the services listed in Exhibit 5–1 should be offered to a sexual assault victim.
- Unit staff should provide more intensive monitoring and supervision to any inmate who has been sexually assaulted. This may include additional meetings of the unit team, application of special monitoring policies, and careful review of custodial and housing assignments. Further, unit staff should refer any inmate who has committed a sexual assault to the mental health staff for evaluation and possible treatment.
- Mental health staff and the program coordinator should offer appropriate care, which may include mental health evaluation and counseling, support services, and follow-up care and tracking. Competency issues of the victim may need to be addressed.
- Chaplaincy staff should offer support and pastoral care, which may include counseling to deal with shame, guilt, and other issues that may surface in the aftermath of an assault.

Exhibit 5–1 Services Available to a Sexual Assault Victim

Medical. Examination, documentation, and treatment of injuries arising from the alleged sexual assault and testing for HIV and other sexually transmitted diseases. The program coordinator should encourage the victim to complete an inmate injury form. If necessary, the victim should be transported to a community emergency facility.

Psychological. Crisis counseling, coping skills, suicide prevention, and mental health counseling as needed. This part of the program may include cooperative arrangements with social service agencies, community-based organizations (such as rape crisis centers), hospitals, or other medical facilities.

Social. Family support, inmate-to-inmate interaction, and peer support groups (if available). The unit and mental health staff should be sensitive to family reactions if the victim notifies friends or relatives of the assault.

Protective. Consultation and staff action to prevent further assaults, including close supervision, protective custody, and possible transfer of the inmate to another facility. If the victim is placed in protective custody or another secure area of the institution, staff members must ensure that the assailant or cohorts of the assailant are not in the area.

- Security and legal staff should coordinate such matters as the collection and corroboration of evidence and witness testimony and consult on administrative and disciplinary issues.

4. Investigation and Prosecution

- Staff members may learn of a sexual assault in one of several ways: the staff may discover an assault in progress; the victim may report an assault to a staff member; an assault may be reported to the staff by another inmate, or may be the subject of inmate rumors that come to the attention of staff; and medical evidence may come to light indicating that an assault has taken place. If any staff member learns that an assault has taken place or may have taken place, then an investigation should be commenced immediately.
- Security staff (including the investigative supervisor, shift commanders, and unit officers) should collect information and evidence to ascertain the facts concerning the alleged assault. Depending on the case and the circumstances, they may work with state police detectives, officers of the correctional agency's Office of Internal Affairs, or other law enforcement representatives on the investigation.
- A brief statement about the assault should be obtained from the victim, if possible. The victim may be in shock and unable to provide many details about the assault, but it is important for staff members to be understanding and responsive at this point in the investigation; opportunities to obtain further details will occur later.

Medical staff should examine the victim to determine whether physical injuries occurred and to render treatment, if appropriate. The medical determination of the degree of bodily harm may be crucial to the investigation. Injuries should be documented in writing and with photographs. To facilitate the collection of medical evidence, the victim should not shower, wash, drink, eat, defecate, or change underwear until examined. When the victim does change into clean clothes, each item of worn clothing should be placed in a separate evidence container. Standard evidence collection procedures (the taking of photographs, etc.) should be observed.

Staff should review the background of the suspected victim and the circumstances surrounding the incident, without jeopardizing the inmate's safety, identity, or privacy.

It is important that all contact with the sexual assault victim be sensitive, supportive, and nonjudgmental; it is not necessary at this point in the investigation to reach a conclusion about whether an assault actually has occurred.

- In the area where the assault allegedly took place, staff members should examine blankets, sheets, and other objects for evidence, including blood and semen. Objects that may have been used as weapons should be secured, and standard investigative techniques should be followed to maintain the chain of evidence.
- If the assailant can be identified, he or she should be cited in an incident report and placed in administrative segregation pending further investigation. The investigation should be handled in accordance with standard disciplinary procedures, as outlined in Chapter 2, Part VIII. Once in administrative segregation, the suspect should be isolated as much as possible and should not be placed with other inmates.
- Staff members should interview any possible witnesses to the assault. Witnesses should be kept apart from the alleged assailant and the alleged victim.
- If, based on the investigation, security staff determine that the assault did take place, and that an assailant can be identified, disciplinary measures against the alleged assailant—up to and including criminal prosecution—may be initiated, in accordance with the correctional agency's disciplinary policies (see Chapter 2, Part VIII). If, based on the investigation, security staff determine that the alleged sexual assault was fabricated, or that the alleged victim inflicted or attempted to inflict self-injury, disciplinary action may be initiated against the inmate claiming to have been assaulted. An assessment of the alleged victim's role and degree of culpability is always necessary, and inmates who report assaults should be informed that such inquiries are routine.

5. Tracking Sexual Assaults

- The primary purpose of the correctional agency's sexual assault and intervention program is to protect inmates in the agency's custody. Monitoring and evaluation are essential to assess both sexual assault levels and agency effectiveness in reducing sexually abusive behavior. Accordingly, the program coordinator should collect and maintain demographic data on each victim and assail-

ant, as well as information on crime characteristics, formal and informal actions taken, services used, and other outcomes of program activities. Data on individual victims should be made available only to staff with a need to know.
- The program coordinator at each institution should submit a brief annual report on their programs to the chief of mental health services at the correctional agency's headquarters. These reports should describe and assess sexual assault data for each institution and provide a qualitative review of program effectiveness. The chief of mental health services will aggregate the findings and prepare a report for review by the deputy director or deputy commissioner in his or her chain of command, who should forward the report on to the director or commissioner.

End Notes

Most of the policies and procedures featured in this volume were adapted from publicly available program statements issued by the Federal Bureau of Prisons, each of which is cited below. Copies of these program statements may be obtained from the Federal Bureau of Prisons. In addition, many of the policies and procedures conform to the *American Correctional Association's Standards for Adult Correctional Institutions* (3rd edition), jointly approved by the American Correctional Association and the Commission on Accreditation for Corrections. Specific standards are cited below and may be obtained from the American Correctional Association.

Chapter 1: The Basics: Processing Inmates Into and out of Prison

I. Receiving and Discharge Operations

Adapted from Federal Bureau of Prisons Program Statement 5800.08, "Receiving and Discharge Manual." Relevant American Correctional Association Standards: 3–4093, 3–4101, 3–4272, 3–4279, and 3–4393.

II. Admission and Orientation Program

Adapted from Federal Bureau of Prisons Program Statement 5290.08, "Admission and Orientation Program." Relevant American Correctional Association Standards: 3–4272, 3–4274, 3–4275, 3–4276, 3–4277, and 3–4278.

III. Pretrial Inmates

Adapted from Federal Bureau of Prisons Program Statement 7331.03, "Pretrial Inmates." Relevant American Correctional Association Standards: 3–4263, 3–4272, 3–4431.

IV. Classification

Adapted from Federal Bureau of Prisons Program Statement 5322.09, "Classification and Program Review of Inmates." Relevant American Correctional Association Standards: 3–4128–1, 3–4273, 3–4276, 3–4278, 3–4282, 3–4283, 3–4284, 3–4285, 3–4286, 3–4287, 3–4288, 3–4289, and 3–4290.

V. Inmate Files

Adapted from Federal Bureau of Prisons Program Statement 5800.09, "Central File, Privacy Folder, and Parole Mini-Files." Relevant American Correctional Association Standards: 3–4092, 3–4095, 3–4096, and 3–4233.

Chapter 2: Day-to-Day Supervision and Security Procedures

I. Inmate Accountability

Adapted from Federal Bureau of Prisons Program Statement 5511.04, "Inmate Accountability." Relevant American Correctional Association Standards: 3–4180, 3–4245, and 3–4181.

II. Searches

Adapted from Federal Bureau of Prisons Program Statement 5521.04, "Searches of Housing Units, Inmates, and Inmate Work Areas." Relevant American Correctional Association Standards: 3–4185, 3–4186, 3–4269.

III. Posted Picture Files

Adapted from Federal Bureau of Prisons Program Statement 5510.05, "Posted Picture Files."

End Notes   241

IV. Alcohol Testing

Adapted from Federal Bureau of Prisons Program Statement 6590.05, "Alcohol Testing."

V. Urine Surveillance

Adapted from Federal Bureau of Prisons Program Statement 6060.05, "Urine Surveillance to Detect and Deter Illegal Drug Use."

VI. The Special Monitoring System

Adapted from Federal Bureau of Prisons Program Statement 5180.03, "Central Inmate Monitoring System."

VII. Special Housing

Adapted from Federal Bureau of Prisons Program Statements 5212.05, "Special Management Unit Operations," and 5270.07, "Special Housing Unit Operations." Relevant American Correctional Association Standards: 3–4214, 3–4215, 3–4216, 3–4217, 3–4218, 3–4220, 3–4221, 3–4223, 3–4226, 3–4227, 3–4228, 3–4229, 3–4230, 3–4231, 3–4232, 3–4233, 3–4234, 3–4236, 3–4237, 3–4238, 3–4239, 3–4240, 3–4241, 3–4242, 3–4243, 3–4244, 3–4246, 3–4248, 3–4249, 3–4250, 3–4251, 3–4253, 3–4254, 3–4255, 3–4256, 3–4257, 3–4258, 3–4259, 3–4260, and 3–4261.

VIII. Inmate Discipline

Adapted from Federal Bureau of Prisons Program Statement 5270.07, "Discipline and Special Housing Units." Relevant American Correctional Association Standards: 3–4214, 3–4215, 3–4216, 3–4217, 3–4218, 3–4219, 3–4220, 3–4221, 3–4222, 3–4223, 3–4226, 3–4227, 3–4228, 3–4229, 3–4230, 3–4231, 3–4232, 3–4233, 3–4234, 3–4236, 3–4237, 3–4238, 3–4240, 3–4243, and 3–4260.

IX. Use of Nonlethal Force and Application of Restraints

Adapted from Federal Bureau of Prisons Program Statements 5558.10, "Custody Control Belt, Use of," and 5566.04, "Use of Force and Application of Restraints on Inmates." Relevant American Correctional Association Standards: 3–4182, 3–4183, 3–4191, 3–4192, 3–4194, 3–4195, and 3–4198.

X. Use of Firearms

Adapted from Federal Bureau of Prisons Program Statement 5558.08, "Firearms and Badges." Relevant American Correctional Association Standards: 3–4081, 3–4088, 3–4197, and 3–4193.

XI. Escorted Trips

Adapted from Federal Bureau of Prisons Program Statement 5538.02, "Escorted Trips." Relevant American Correctional Association Standards: 3–4350, 3–4360, and 3–4392.

XII. Voluntary Surrenders and Unescorted Transfers

Adapted from Federal Bureau of Prisons Program Statement 5140.21, "Unescorted Transfers and Voluntary Surrenders."

XIII. Security Considerations Relating to Inmate Access to Computers

Adapted from Federal Bureau of Prisons Program Statement 1237.07, "Computer Security."

Chapter 3: Inmate Entitlements

I. Inmate Telephone Regulations

Adapted from Federal Bureau of Prisons Program Statement 5264.05, "Telephone Regulations for Inmates."

II. Inmate Correspondence

Adapted from Federal Bureau of Prisons Program Statements 5265.08, "Correspondence," and 5266.05, "Incoming Publications." Relevant American Correctional Association Standards: 3–4254, 3–4430, 3–4431, 3–4432, 3–4433, 3–4435, 3–4436, and 3–4438.

III. Visiting Regulations

Adapted from Federal Bureau of Prisons Program Statement 5267.05, "Visiting Regulations." Relevant American Correctional Association Standards: 3-4255, 3-4272, 3-4440, 3-4441, 3-4442, 3-4445, 3-4446.

IV. Inmate Grievance Procedures

Adapted from Federal Bureau of Prisons Program Statement 1330.11, "Administrative Remedy Procedure for Inmates." Relevant American Correctional Association Standards: 3-4236, 3-4271, 3-4393, 3-4434.

V. Inmate Legal Activities

Adapted from Federal Bureau of Prisons Program Statement 1315.05, "Legal Activities, Inmate." Relevant American Correctional Association Standards: 3-4262, 3-4263, 3-4264, and 3-4442.

VI. News Media Contacts

Adapted from Federal Bureau of Prisons Program Statement 1480.03, "Contact with News Media." Relevant American Correctional Association Standards: 3-4021, 3-4022, and 3-4267.

Chapter 4: Programs and Services

I. Inmate Employment, Educational Programs, and Vocational Training

Adapted from Federal Bureau of Prisons Program Statements 5300.15, "Education, Training, and Leisure Time Program Standards," and 5300.16, "Volunteers and Citizen Participation Programs Manual. Relevant American Correctional Association Standards: 3-4111, 3-4112, 3-4113, 3-4114, 3-4115, 3-4116, 3-4117, 3-4118, 3-4119, 3-4264, 3-4395, 3-4401, 3-4412, 3-4414, 3-4415, 3-4416, 3-4418, 3-4419, 3-4421, 3-4422, 3-4426, and 3-4428.

II. Recreational Activities

Adapted from Federal Bureau of Prisons Program Statement 5370.08, "Recreation Programs, Inmate." Relevant American Correctional Association Standards: 3–4423, 3–4424, 3–4425, 3–4426, 3–4427, and 3–4428.

III. Religious Activities

Adapted from Federal Bureau of Prisons Program Statements 5360.05, "Religious Beliefs and Practices of Committed Offenders," and 5300.16, "Volunteers and Citizen Participation Programs Manual." Relevant American Correctional Association Standards: 3–4111, 3–4112, 3–4113, 3–4114, 3–4115, 3–4116, 3–4117, 2–4118, 3–4119, 3–4300, 3–4395, 3–4455, 3–4456, 3–4457, 3–4458, 3–4459, and 3–4462.

IV. Inmate Organizations

Adapted from Federal Bureau of Prisons Program Statement 5381.03, "Inmate Organizations."

V. Halfway House Placement

Adapted from Federal Bureau of Prisons Program Statement 7310.02, "Community Corrections Centers Utilization and Transfer Procedure." Relevant American Correctional Association Standards: 3–4391 and 3–4393.

VI. Furloughs

Adapted from Federal Bureau of Prisons Program Statement 5280.06, "Furloughs." Relevant American Correctional Association Standards: 3–4390, 3–4443, and 3–4444.

Chapter 5: Medical and Psychological Issues

I. Infectious Disease Management

Adapted from Federal Bureau of Prisons Program Statements 5214.03, "HIV, Handling of Inmates Testing Positive," and 6190.01, "Human Immunodeficiency Virus." Relevant American Correctional Association Standards: 3–4268, 3–4365, and 3–4366.

II. Suicide Prevention Program

Adapted from Federal Bureau of Prisons Program Statement 5324.01, "Suicide Prevention Program." Relevant American Correctional Association Standards: 3-4081, 3-4343, 3-4245, and 3-4364.

III. Food from Outside Sources

Adapted from Federal Bureau of Prisons Program Statement 4761.03, "Special Food or Meals from Outside Sources Introduced into Institutions."

IV. Inmate Hunger Strikes

Adapted from Federal Bureau of Prisons Program Statement 5562.04, "Hunger Strikes." Relevant American Correctional Association Standards: 3-4245 and 3-4372.

V. Sexual Assault Prevention and Intervention

Adapted from Federal Bureau of Prisons Program Statement 5324.02, "Sexual Assault Prevention/Intervention Program, Inmate." Relevant American Correctional Association Standards: 3-4081 and 3-4268.

INDEX

A

Academic counseling, 176
Acknowledgment forms, inmate processing, 7–8
Administrative segregation units, 73
  inmate placement, 76–78
  reasons for confinement, 76
  time factors, 77–78
  See also Special housing units
Admission of inmates. *See* Intake process
Admission/orientation program, 17–20
  documentation, 19–20
  elements of, 18
  eligibility for, 17
  location for, 18–19
  purpose of, 17
  schedule for, 18
  staff roles, 19
  time frame for, 18–19
Adult continuing education activities, 175
Advance authorization form, mail/packages, 9
Alcohol testing, 61–62
  of liquids, 62
  procedures for, 61–62
Ambulatory restraints, 108
Appeals
  of classification, 36
  disciplinary actions, 104–105
  of grievance decision, 159
  special monitoring classification, 71–72
Apprenticeship training, 175
Art, inmate works, 183–184
Athletic activities, 182–183
Attorney
  confidential communication and inmate, 138
  telephone calls to, 138
Authorized Unescorted Commitment and Transfer form, 198

B

Bilingual programs, 178
Body search, 53–54
  digital search, 54
  metal detector searches, 53
  pat search, 53
  simple instrument search, 54
  visual search, 53–54

C

Call out sheet, 51
Career counseling, 176
Case manager, role of, 31, 33, 81
Certificates, from educational programs, 177–178
Check-out cards, inmate files, 43–44
Chemical agents, nonlethal force, 112
Chief medical officer, role of, 206–207,

208–209, 210–211
Children, visiting, 153
Classification, 30–36
　appeals procedure, 36
　detainers, 36
　initial classification, 32
　program reviews, 32–33, 35–36
　purpose of, 30
　special monitoring classification, 67–73
　special monitoring procedures, 68–70
　study and observation cases, 36
Classification team, 30–31, 33–34
　case manager, 31, 33
　correctional counselor, 31, 33–34
　education advisor, 30–31, 33
　staff training, 36
　unit manager, 31, 32, 33, 36
　unit psychologist, 31, 34
Clearance, special monitoring classification, 72–73
Collect calls, 137, 138
Commissary account
　close-out of, 16
　forms for new inmates, 8
　funds from outside prison, 148
Commitment documentation
　examples of, 3
　intake processing, 2–4
Committed Fine Form, 127
Community corrections
　community custody, requirements for, 124
　forms of, 193
　halfway house, 193–194
　release plan, 194
　subsistence charge, 194
　transfer to facilities, 124–125, 126–127
Comprehensive Adult Student Assessment System, 174
Computer Security Committee, 128
Computers, inmate use, 128–132
　allowable activities, 128–129
　disks, control of, 131
　prohibited activities, 129–130
　prohibited inmates, 132
　special precautions, 130–132
Conditions of Supervised Release Form, 127
Confidential informants

and disciplinary actions, 103–104
special monitoring classification, 69, 71
Confrontation with inmates
　avoidance procedures, 109–110
　firearms, 115–118
　nonlethal force, 105–115
　restraints, use of, 106–109
Contraband, removal in search, 54, 58
Control center records, contents of, 49
Corporal punishment, 89
Correctional counselor, role of, 31, 33–34, 81
Correspondence, 138–149
　general correspondence, 139–142
　high-risk inmates, 141
　and inmate classification, 140–141
　and inmate transfers, 149
　inspection of, 140–141
　legal mail, 143–144
　open general correspondence, 139
　postage, 146–147
　procedures for, 147–149
　publications, 144–145
　rejected correspondence, 141
　restricted general correspondence, 139–140
　special mail, 142–143
Counseling, for pretrial inmates, 28
Court orders, intake processing, 3
Criminal Law Reporter, 161
Custody control belt, 113–115
　guidelines for use, 113–115
Custody level
　community custody, 124
　in custody, 124
　maximum custody, 124
　out custody, 124
　pretrial inmates, 23–24

D

Deportation, pretrial inmates, 24
Detail accountability check, 50–51
Detainers, classification, 36
Diet
　religious diets, 187–188
　See also Meals
Digital search, 54

Discharge of inmates. *See* Out-processing inmates
Disciplinary detention units, 73
  inmate placement, 75–76
  inmate release, 76
  reasons for confinement, 75
  See also Special housing units
Disciplinary hearing, 97–99
  actions resulting from, 98, 101–102
  disciplinary hearing officer, role of, 100–102
  documentation, 99
  unit discipline committee, 95, 97–100
Discipline, 88–105
  for acts of greatest severity, 90–91
  for acts of high severity, 91–92
  for acts of low severity, 93–94
  for acts of moderate severity, 92–93
  appeals, 104–105
  basic principles, 88–89
  confidential informants, 103–104
  corporal punishment, 89
  disciplinary hearing officer, 100–102
  hearing for inmate, 97–99
  incident reports, 94–95
  investigation for, 95–97
  and mentally ill inmate, 89
  meting out sanctions, 94
  notification of inmates, 94
  pretrial inmates, 26
  unit discipline committee, role of, 97–100
Disease prevention
  levels of, 206
  See also Medical examinations; Medical programs
Documentation
  admission/orientation program, 19–20
  disciplinary hearing, 99
  nonlethal force, 110–111, 112
  nonlethal force and restraints, 107–108
  out-processing inmates, 14–15
  procedure in, 15–6
  special housing unit placement, 75
Drug screening
  positive results, actions for, 66
  urine testing, 63–67
Dry cells, requirements for, 55

Dry cell status, 54–57
  requirements for inmates, 56–57

E

Education advisor, role of, 30–31, 33, 81, 85
Educational programs
  academic/career counseling, 176
  attendance monitoring, 176
  bilingual programs, 178
  certificates from, 177–178
  English as a Second Language, 174
  for General Equivalency Diploma (GED), 173–174
  handbook for, 179
  inmate completion of, 177
  instructors, 179–180
  literacy programs, 173–174
  occupational education, 174–175, 179
  pretrial inmate participation, 29
  progress/evaluation report, 178
  records, 176–178
  schedule for, 179
  and special housing inmates, 85
  standards, 177
  transcripts, 178
  volunteers, 180
Emergency trips
  medical, 118–120
  nonmedical, 120–121
English as a Second Language, 174
Entitlements to inmates
  correspondence, 138–149
  grievance procedures, 156–160
  legal materials, access to, 160–166
  media contact, 166–169
  telephone use, 133–138
  visitors, 149–156
Entrance pass, front/rear entrance pass, 52
Environmental controls, medical, 206
Enzyme-linked immunosorbent assay (ELISA), 216
Escape, use of firearms, 116
Escorted Trip Authorization Form, 14–15, 114, 118, 120, 123
Escorted trips, 118–123
  custody control belt, use of, 113–115

emergency trips, 120–121
escort procedures, 122–123
escort staff selection, 122
firearms for staff escorts, 116
medical trips, 118–120
for special monitoring inmates, 121–122
Evidence of Agent's Authority to Act for Receiving State Form, 17
Exploratory training, occupational, 174–175

F

FBI Arrest Fingerprint Card, 5
Federal prisoners, special monitoring classification, 69, 71
Files. *See* Inmate files
Fingerprints, inmate processing, 4–6
Firearms, 115–118
　conditions for carrying firearms, 115–116
　discharge, situations for, 116–117
　safeguards, 117–118
　staff training, 117–118
Fluoroscope examination, 57
Force
　avoidance procedures, 109–110
　firearms, 115–118
　nonlethal force, 105–115
　restraints, 106–109
Four-point restraints, 108–109
Fund raising, inmate organizations, 190
Furlough, 198–204
　approval procedure, 202–204
　criteria for authorization, 200–201
　definition of, 198–199
　duration of, 199
　eligibility criteria, 201–202
　expenses, 199–200
　and pretrial inmates, 28
　and public information about program, 204
　transfers. *See* Unescorted transfers
　violation of conditions, 204
Furlough Application and Approval Form, 14, 127

G

General Equivalency Diploma (GED), 173–174
Government threats, special monitoring classification, 69
Greatest severity, discipline category, 90–91
Grievance procedures, 156–160
　appeal of decision, 159
　filing complaint, 158–159
　forms/paperwork, 157–158
　informal resolution, 158
　rejected cases, 156
　review of filings, 157

H

Halfway house, 192–198
　criteria for placement, 195–196
　inmate refusal of placement, 196
　referral procedures, 196–198
　services/operation of, 193–194
　unescorted transfer, 126–127
Handcuffs. *See* Restraints
Head count system, 49
Health administrator, role of, 206–207
Health promotion programs, 205–206
　types of, 206
Health protection, meaning of, 206
Hepatitis B testing, 214–217
High severity, discipline category, 91–92
HIV-positive inmates, 217–219
　housing, 211, 217, 218–219
HIV testing, 208, 214–217
　tests used, 216
Hobbies, of inmates, 183–184
Holding cells, receiving/discharge area, 12–13
Holidays, religious, 188–189
Hot trash, 10
Housing units
　pretrial inmates, 24
　searches, 58
　special housing units, 73–88
Hunger strikes, 227–231
　forced feeding, 229–230
　medical management, 228–229
　medical referral, 228

release from treatment, 230–231
supply of food/water to inmate, 229

I

Incident reports, for disciplinary actions, 94–95
In custody, requirements for, 124
Industrial work assignments, 172–173
Infectious disease training, 209–210
Informants. *See* Confidential informants
Inmate files, 37–48
  check-out cards, 43–44
  core documents, 42
  creation of files, 38
  education records, 176–179
  inactive files, retirement of, 48
  judgment and commitment files, 41
  laws/regulations related to, 37
  location/storage of, 42–43
  maintenance of, 43–44
  parole board files, 39, 41
  privacy folder, 39
  records schedule, 37
  responsibility for, 37–38
  review by inmates, 46–47
  sections for, 40–41
  security, 43–44
  transfer of records, 47–48
  uses of, 44–46
Inmate organizations, 189–192
  expenses, 190–191
  financial reports, 191–192
  fund raising, 190
  requirements/criteria, 189–190
Instructors, educational programs, 3–4
Intake process, 1–12
  acknowledgment forms, 7–8
  admission/orientation program, 17–20
  basic procedures, 1
  classification, 30–36
  commissary forms, 8
  fingerprints, 4–6
  ID photos, 4
  inmate files, 37–48
  inmate property processing, 8–12
  medical screening, 7, 27, 214
  pretrial inmates, 20–30, 21

registration numbers, 6–7
search for contraband, 9–10
shower/clothing, 7
In-Transit Information Form, 17
Investigation, for disciplinary actions, 95–97

J

Judgment and Commitment Order, 41
  intake processing, 3
  originals on file, 41

L

Law library, for special management inmates, 87
Legal aid programs, inmate access to, 166
Legal counsel
  denial of visiting privileges, 165
  legal aid programs, 166
  retention by inmates, 164–166
  telephone calls to attorneys, 138
  visits by, 164–165
Legal materials, access to, 160–166
  coping materials, 162, 163–164
  and inmate processing, 9
  law libraries, 160–162
  legal mail, 143–144
  for pretrial inmates, 27
  retaining materials in living quarters, 162, 163
  special housing inmates, 86–87
  time/schedule for, 162
Leg irons. *See* Restraints
Letters/mail. *See* Correspondence
Library, law library, 160–162
Literacy programs, 173–174
Lock-down accountability check, 50
Low severity, discipline category, 93–94

M

Mail/packages
  advance authorization form, 9
  See also Correspondence

Marketable skill training, 175
Marriage request, pretrial inmates, 29
Maximum custody, requirements for, 124
Meals
   holiday/banquet meals, 226–227
   outside food, inspection of, 227
   religious diets, 187–188
Media contact, 166–169
   inmate interviews, 167–168
   location for, 168
   press pool, 168–169
   rejection by warden, 168
   release of information, 169
Medical examinations
   confidentiality, 211–212
   and custody control belt, 114, 115
   escorted medical trips, 118–120
   general medical management, 206–207
   hepatitis B testing, 214–217
   and HIV-positive inmates, 217–219
   HIV testing, 208, 214–217
   inmate processing, 7, 23, 27, 214
   isolation/quarantine, 210–211
   medical testing, 207–209
   special housing inmates, 86
   tuberculosis control, 212–214
Medical programs
   health promotion programs, 205–206
   infectious disease training, 209–210
   OSHA standards, 210
   public health information sources, 210, 212
   wellness program, 185
Mentally ill inmate, discipline, 89
Metal detector searches, 53
Moderate severity, discipline category, 92–93
Movie viewing, 182

N

National Crime Information Center (NCIC), 4
Nonlethal force, 105–115
   calculated use, 106, 110
   chemical agents, 112
   custody control belt, 113–115
   documentation, 107–108, 110–111, 112
   guidelines for use, 106
   immediate use, 105–106
   nonlethal weapons, 112
   safeguards for team, 112–113
   use of force team, role of, 111
   See also Firearms; Restraints
Notice of Separation Form, 21
Notice of Special Management Unit Hearing, 79

O

Occupational education, 174–175
   trade advisory committees, 179
   types of programs, 174–175
Open general correspondence, 139
Open house, receiving/discharge area, 13
Organizations. *See* Inmate organizations
Out custody, requirements for, 124
Out-of-state prisoners, special monitoring classification, 69, 71
Out-processing inmates, 14–17
   documentation, 14–15
   funds for, 16
   property of inmates, 16

P

Parole board files, 39, 41
Parole board warrants, intake processing, 3
Pass system, 51–52
Pat search, 53
Pepper mace, 112
Photographs, 21
   ID, inmate processing, 4
Postage, 146–147
Posted picture file, 58–61
   inmate criteria, 59
   responsibility for, 60
   reviews of, 60–61
Postsecondary education activities, 175
Prerelease training, 175
Pretrial inmates, 20–30
   access to legal materials, 27

admission procedures, 21
assessment information sources, 22–23
compared to convicted inmates, 20
counseling, 28
custody level, 23–24
discipline, 26
housing assignments, 24
intake screening, 21–23
program activities, 28–29
property disposition, 25–26
status reviews, 24–25
visiting privileges, 29–30
Privacy folder, inmate files, 39
Processing inmates
  incoming inmates, 1–12
  out-processing inmates, 14–17
  receiving/discharge area layout, 12–14
Program activities, special housing inmates, 85
Program reviews, 35–36
  for classification, 32–33
  report, 34–35
  time frame for, 32
  unscheduled reviews, 36
Property of inmates
  admission processing, 8–12
  out-processing, 16
  pretrial inmates, 25–26
  special housing units, 84–85
  storage of, 10–11
Protection of inmates
  in administrative segregation units, 76–77
  from sexual assault, 232–233
Publications, incoming mail, 144–145
Public service assignments, 172

Q

Quarantine, medical reasons, 210–211, 218

R

Receiving/discharge area
  layout of, 12–14
  open house, 13

Recreational activities, 180–185
  arts/hobby crafts, 183–184
  athletic activities, 182–183
  at-risk inmates, 181
  inmate needs, 180
  movies/television, 182
  pretrial inmate participation, 29
  schedule, 181
  special housing inmates, 83–84
Recreation supervisor, role of, 180–181
Registration numbers, inmate processing, 6–7
Release of inmate, property of inmate, treatment of, 11
Release plan, community corrections, 194
Religious activities
  guidelines, 185–186
  holidays/celebrations, 188–189
  personal religious items, 186–187
  pretrial inmate participation, 28–29
  special housing inmates, 85–86
Religious beliefs
  religious diets, 187–188
  and work assignments, 188
Religious Freedom Restoration Act (RFRA), 185
Remand to Custody Form, 21
Remand of Prisoner Forms, intake processing, 3
Reports
  program reviews, 34–35
  urine testing, 66–67
Restraints
  ambulatory restraints, 108
  counterindications for use, 107
  documentation, 107–108
  four-point restraints, 108–109
  guidelines for use, 106–107
Restricted general correspondence, 139–140
Review
  program reviews, 35–36
  segregation review, 77
  status review, pretrial inmates, 24–25

S

Searches, 53–58

body search, 53–54
dry cell status, 54–57
housing units, 58
inmate processing, 9–10
pretrial inmates, 21
removal of contraband, 54, 58
special housing inmates, 88
surgical intrusion searches, 57–58
technique for, 10
work areas, 58
X-ray examination, 57–58
Security threats, special monitoring classification, 69
Segregation of inmates. *See* Special housing units
Segregation review, 77
Segregation review officer, role of, 77, 105
Separation cases, special monitoring classification, 69
Sexual assault prevention
  intervention for victims, 233–235
  investigation of incident, 235–236
  monitoring activities, 236–237
  prevention efforts, 232–233
  program coordinator, role of, 231–232
  services for victim, 234
  staff training, 232–233
Shower/clothing, inmate processing, 7, 21
Simple instrument search, 54
Sophisticated criminals, special monitoring classification, 69
Special activities, types of, 192
Special housing units, 73–88
  administrative segregation units, 73, 76–78
  conditions of confinement, 82
  disciplinary detention units, 73, 75–76
  documentation, 75
  duration of confinement, 87
  educational activities, 85
  health care, 86
  legal materials, 86–87
  living conditions, 82–83
  program activities, 85
  property of inmates, 84–85
  provisions to inmate, 83
  recreational activities, 83–84
  religious services, 85–86

searches, 88
segregation review, 77
special management units, 73–74, 78–82
staff responsibilities, 74–75
suicide risk and inmates, 224–225
visitors, 155–156
work assignments, 85
Special mail, 142–143
Special management units, 73–74
  access to legal materials, 160
  final decision, 80–81
  inmate placement, 78–82
  pre-placement hearing, 79–80
  reasons for confinement, 78
  team members for, 81
  See also Special housing units
Special monitoring cases, 67–73
  classification, 68–70
  classification appeals, 71–72
  clearance for inmate, 72–73
  escorted trips, 121–122
  inmate criteria, 69
  purpose of, 67–68
  removal of inmate, 70, 71
  responsibilities/roles, 68
  status review, 71
Special supervision, special monitoring classification, 69
Standards
  educational programs, 177
  infection management, 210
Status reviews, pretrial inmates, 24–25
Strip search, 53–54
Study and observation cases, classification, 36
Stun guns, 112
Subsistence Agreement Form, 198
Subsistence charge, community corrections, 194
Suicide prevention, 219–226
  analysis of suicides, 226
  high-risk inmates, identification of, 220–221, 224–225
  intervention, 221–224
  and special housing units, 224–225
  staff training, 220
  transfer of inmate, 225–226
Suicide program coordinator, role of, 219–220

Supervision of inmates
　alcohol testing, 61–62
　change and transfer sheet, 51
　computers, inmate access to, 128–132
　control center records, 49
　detail accountability check, 50–51
　escorted trips, 118–123
　firearms, 115–118
　front/rear entrance pass, 52
　head count system, 49
　inmate call-outs, 51
　lock-down accountability check, 50
　nonlethal force, 105–115
　pass system, 51–52
　posted picture file, 58–61
　restraints, 106–109
　searches, 53–58
　special accountability, 52
　special housing units, 73–88
　special monitoring cases, 67–73
　unescorted transfer, 125–127
　urine testing, 62–67
　voluntary surrenders, 123–125
　work crew kit cards, 52
Surgical intrusion searches, 57–58

T

Telephone, 133–138
　calls to attorneys, 138
　collect calls, 137, 138
　expenses to inmate, 136–137
　limitation of use, 133
　monitoring of calls, 137–138
　official lists for, 133–135
　restricted persons, 135
Television viewing, 182
Trade advisory committees, 179
Transcripts, from educational programs, 178
Transfer Orders, intake processing, 3
Transfers
　and correspondence, 149
　to halfway house, 126–127
　of records, 47–48
　of suicidal inmates, 225–226
　unescorted transfer, 125–127
　voluntary surrender, 124–125
Transfer sheet, 51

Tuberculosis, control guidelines, 212–214

U

Unescorted initial commitments. *See* Voluntary surrenders
Unescorted transfer, 125–127
　to another prison, 125–126
　to halfway house, 126–127
Unit discipline committee, role of, 95, 97–100, 103, 104
Unit manager, role of, 31, 33, 34, 36, 81
Unit psychologist, role of, 31, 34
Urine testing, 62–67
　drug screening, 66
　inmate criteria, 63–64
　positive test, actions for, 65–66
　reports, 66–67
　sampling procedures, 64–66
　staff roles, 64–65
　testing laboratory, 66
Use of force team
　roles of, 111
　safeguards for team, 112–113

V

Victim Witness Notification, 127
Visitors, 149–156
　attorneys, 164–165
　children, 153
　and inmate classification, 151
　for pretrial inmates, 29–30
　regular visitors, 151
　schedules, 153–154
　special approval, 152
　for special management inmates, 87
　for special status inmates, 155–156
　special visitors, 151–152
　supervision, 154
　visiting areas, 153
　visitor lists, 149–150
Visual search, 53–54
Voluntary surrenders, 123–125
　guidelines for, 125
Volunteers, educational programs, 180

W

Warden, and administrative segregation of inmates, 76
Wellness program, 185
Western Blot, 216
Witnesses, special monitoring classification, 69, 71
Work, 171–173
  assignments and special housing inmates, 85
  commissary account, inmate earnings, 172
  forms of employment, 172–173
  inmate earnings, 171–172
  supervision of inmates, 173
  work areas, searches, 58
Work crew kit cards, 52
Writ Returns, intake processing, 3
Written notification, of disciplinary actions, 94

X

X-ray examination, 57–58

ABOUT THE AUTHORS

Richard L. Phillips is an experienced correctional manager who has worked in juvenile and adult corrections at both state and federal levels. His 30 years of correctional experience includes field management assignments in minimum, medium, and high security felony facilities, as well as in an urban detention setting, an agency regional office, and headquarters administrative post. He is co-author of *The Effective Corrections Manager: Maximizing Staff Performance in Demanding Times* (Aspen Publishers, 1996), and *Guidelines for the Development of a Security Program* (American Correctional Association, 1997).

John W. Roberts is a Senior Archivist with History Associates, Inc., in Rockville, Maryland, with more than 10 years experience in the corrections field. His books include *Reform and Retribution: An Illustrated History of American Prisons* (American Correctional Association, 1997), as well as *Escaping Prison Myths: Selected Topics in the History of Federal Corrections* (American University Press, 1994). He has also contributed to periodicals, including *Corrections Today*, and *Federal Prisons Journal*.